Graphing Calculator Supplement For Texas Instruments Calculator

Carolyn L. Meitler

Finite Mathematics

for Business, Economics, Life Sciences, and Social Sciences

Eighth Edition

Raymond A. Barnett

Michael R. Ziegler

Karl E. Byleen

PRENTICE HALL, Upper Saddle River, NJ 07458

Executive Editor: Sally Simpson
Editorial Assistant: Sara Beth Newell
Special Projects Manager: Barbara A. Murray
Production Editor: Michele Wells
Supplement Cover Manager: Paul Gourhan
Supplement Cover Designer: Liz Nemeth
Manufacturing Buyer: Alan Fischer

Simon & Schuster / A Viacom Company
Upper Saddle River, NJ 07458

Printed in the United States of America

10 9 8 7 6 5 4 3 2 1

ISBN 0-13-932195-0

Prentice-Hall International (UK) Limited, *London*
Prentice-Hall of Australia Pty. Limited, *Sydney*
Prentice-Hall Canada, Inc., *London*
Prentice-Hall Hispanoamericana, S.A., *Mexico*
Prentice-Hall of India Private Limited, *New Delhi*
Prentice-Hall of Japan, Inc., *Tokyo*
Simon & Schuster Asia Pte. Ltd., *Singapore*
Editora Prentice-Hall do Brazil, Ltda., *Rio de Janeiro*

Table of Contents

PREFACE

Graphing calculators are powerful tools that can help in learning mathematics. This manual includes examples that illustrate how to use the calculator with topics covered in a finite mathematcs course. A detailed solution to each example shows how a graphing calculator can be used to solve the problem.

Appendices

Four appendices are included. Each appendix contains keystrokes necessary to solve problems for the following ™Texas Instruments calculators:

Appendix A	TI-82 Operations
Appendix B	TI-83 Operations
Appendix C	TI-85 Operations
Appendix D	TI-86 Operations

More than one method of solution using the calculator is shown for many examples. Neither programming nor parametric equations are used.

The appendices correspond to one another. The same example is used in each appendix. This allows instruction to be given to a class of students in which more than one calculator model is being used.

Calculations

Default settings on the calculator have been used for all calculations in the problem solutions, unless otherwise noted. Numbers are rounded to the desired number of significant digits or number of decimal places as the problem requires after all calculations have been completed. Directions on how to set the calculator to a specified number of significant digits, or a given number of decimal places, before performing the calculations are given in Appendices A-17, B-17, C-17 and D-17. Setting a specified number of decimal places was not done in the examples of this manual but might be convenient in many problems.

Built-in Calculator Functions

The built-in functions (features) of the calculator are used in some of the examples. Discussion of some of the built-in features for a particular calculator can also be found in the appendix pertaining to that calculator.

Graphs

The curves shown in the figures of this manual are not smooth but rather are intended to represent the graphing calculator display.

ACKNOWLEDGMENTS

The completion of this project would not have occurred without the support and assistance of several individuals. First, and foremost, is Michael R. Ziegler. His suggestions were most helpful in forming the philosophy, layout, and content of this manual. Also I would like to thank James Briselden who helped in checking for accuracy and consistency. Finally, I thank my family for their encouragement and support.

Carolyn L. Meitler

Chapter 1

A Beginning Library of Elementary Functions

This chapter contains examples using the calculator to illustrate:

- Plotting points and drawing lines
- Graphing functions
- Graphing piecewise-defined functions
- Evaluating algebraic expressions
- Determining the domain of a function from a graph
- Evaluating functions for a given value of x
- Investigating the change in the value of an expression as the value of a variable changes
- Investigating the change in the graph of a function when constant terms are changed
- Finding intercepts
- Solving an equation in one variable
- Solving an inequality in one variable

EXAMPLE 1 ***TEXTBOOK SECTION 1-1***

(A) Plot the points (-5, 2) and (1, 3). (B) Draw a line segment with these points as the endpoints.

Solution (A): First viewing window variables need to be set. For this example, set Xmin=-10, Xmax=10, Xscl=1, Ymin=-10, Ymax=10, and Yscl=1.

Get the [Pt-On(] function on the calculator. Enter the values -5 and 2 and plot the first point. Get the [Pt-On(] function again and plot the second point. (See Appendix Section A-3, B-3, C-3 or D-3 of this manual.)

Solution (B): Get the [Line(] function of the calculator and draw the line. (See Appendix Section A-3, B-3, C-3 or D-3 of this manual.)

EXAMPLE 2 ***TEXTBOOK SECTION 1-1***

(A) Find the y coordinate of the points on the graph of $y = 3 - x^2$ for x = -3, -1, 0, 1, 3. (B) Plot the resulting ordered pairs. (C) Graph the function.

Solution (A): Store $3 - x^2$ in the function list in your calculator as: 3 - X ^ 2. (See Appendix Section A-4, B-4, C-4 or D-4.) One method to obtain the y coordinates is to store the first value for x in the calculator and evaluate the function. Repeat for the other four values of x. The resulting ordered pairs are (-3, -6), (-1, 2), (0, 3), (1, 2), and (3, -6).

A second method is to store the values for x as a list and to evaluate the function for the entire list. First store $3 - x^2$ in the function list. Then store the list and evaluate it.

TI-82

[2nd] [{] [(-)] [3] [,] [(-)] [1] [,] [0] [,] [1] [,] [3] [2nd] [}] [STO▸] [2nd] [L1] [ENTER]
[2nd] [Y-VARS] [1] <Function...> [1] <Y1> [(] [2nd] [L1] [)] [ENTER]

TI-83

[2nd] [{] [(-)] [3] [,] [(-)] [1] [,] [0] [,] [1] [,] [3] [2nd] [}] [STO▸] [2nd] [L1] [ENTER]
[VARS] [▶] [1] <Function...> [1] <Y1> [(] [2nd] [L1] [)] [ENTER]

TI-85 & TI-86

[2nd] [LIST] [F1] <{> [(-)] [3] [,] [(-)]
[1] [,] [0] [,] [1] [,] [3] [F2] <}> [STO▸] [x-VAR] [ENTER]
[2nd] [ALPHA] [y] [1] [ENTER]

Solution (B): Refer to Example 1 above on how to plot points on the calculator. Set the viewing window variables to [-10, 10]1 by [-10, 10]1 (see Appendix Section A-3, B-3, C-3 or D-3 of this manual).

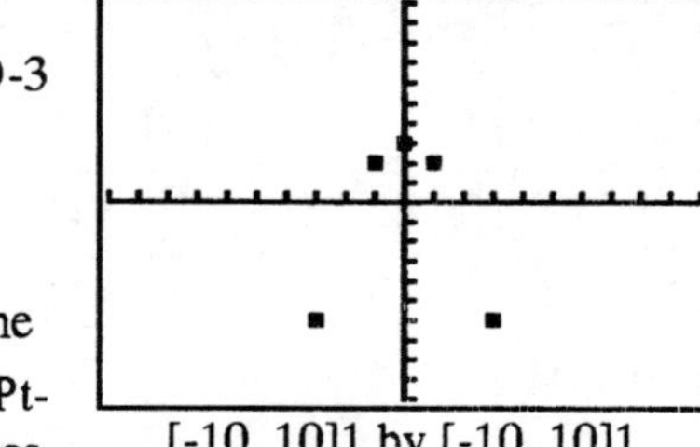

[-10, 10]1 by [-10, 10]1

Method 1 Plot each point separately

Get the [Pt-On(] function on the calculator. Enter the values -3 and -6 and plot the first point. Get the \x(Pt-On() function again, enter the values of the coordinates to plot the second point. Repeat for the other points. The result is shown to the right. (See Example 1 above and Appendix Section A-3, B-3, C-3, or D-3 of this manual.)

Method 2 Plot all the points at one time

Store the points in the lists for the statistics function of the calculator. Draw a scatter diagram. (See Appendix Section A-16, B-16, C-16 or D-16 of this manual.) Be sure to clear or deselect all functions in the function list. (See Appendix Section A-3, B-3, C-3 or C-3 of this manual.)

TI-82

[STAT] [4] <ClrList> [2nd] [L1] [,] [2nd] [L2] [ENTER] [STAT]
[1] <Edit…> [(-)] [3] [ENTER] [(-)] [1] [ENTER] [0] [ENTER] [1] [ENTER] [3] [ENTER]
[▶] [(-)] [6] [ENTER] [2] [ENTER] [3] [ENTER] [2] [ENTER] [(-)] [6] [ENTER]
[2nd] [STAT PLOT] [1] [ENTER] [▼] [ENTER] [▼] [ENTER] [▼] [▶] [ENTER] [GRAPH]

TI-83

[STAT] [4] <ClrList> [ENTER] [2nd] [L1] [,] [2nd] [L2] [ENTER]
[STAT] [1] <Edit…> [(-)] [3] [ENTER] [(-)] [1] [ENTER] [0] [ENTER] [1] [ENTER]
[3] [ENTER] [▶] [(-)] [6] [ENTER] [2] [ENTER] [3] [ENTER] [2] [ENTER] [(-)] [6]
[ENTER] [2nd] [STAT PLOT] [1] [ENTER] [▼] [ENTER] [▼] [2nd] [L1] [▼]
[2nd] [L2] [GRAPH]

TI-85

[STAT] [F2] <EDIT> [F1] <xStat> [ENTER] [F2] <yStat> [ENTER] [F5] <CLRxy>
[(-)] [3] [ENTER] [(-)] [6] [ENTER] [(-)] [1] [ENTER] [2] [ENTER]
[0] [ENTER] [3] [ENTER] [1] [ENTER] [2] [ENTER] [3] [ENTER]
[(-)] [6] [EXIT] [F3] <DRAW> [F2] <SCAT> [CLEAR]

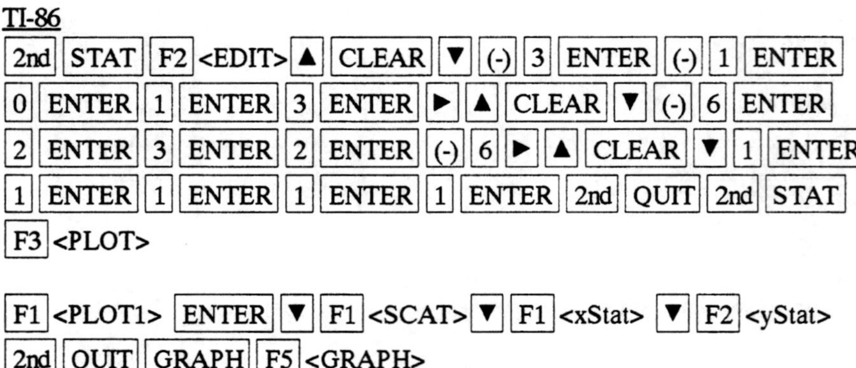

TI-86

2nd STAT F2 <EDIT> ▲ CLEAR ▼ (-) 3 ENTER (-) 1 ENTER
0 ENTER 1 ENTER 3 ENTER ► ▲ CLEAR ▼ (-) 6 ENTER
2 ENTER 3 ENTER 2 ENTER (-) 6 ► ▲ CLEAR ▼ 1 ENTER
1 ENTER 1 ENTER 1 ENTER 1 ENTER 2nd QUIT 2nd STAT
F3 <PLOT>

F1 <PLOT1> ENTER ▼ F1 <SCAT> ▼ F1 <xStat> ▼ F2 <yStat>
2nd QUIT GRAPH F5 <GRAPH>

Solution (C): Set the viewing window variables to [-10,10]1 by [-10, 10]1 (see Appendix Section A-6, B-6, C-6, or D-6 of this manual). Store $3 - x^2$ in the function list in your calculator if it has not already been stored. Select the function if it has been deselected.

Clear or deselect all other functions. (See Appendix Section A-6, B-6, C-6, or D-6 of this manual.)

Clear all drawings. (See Appendix Section A-3, B-3, C-3, or D-3.)

Turn all statistical plots off if you are using the TI-82, TI-83 or TI-86 calculators. (See below and Appendix Section A-16, B-16, or D-16 of this manual.) If you do not turn the statistical plots off the points will plot as well as the function.

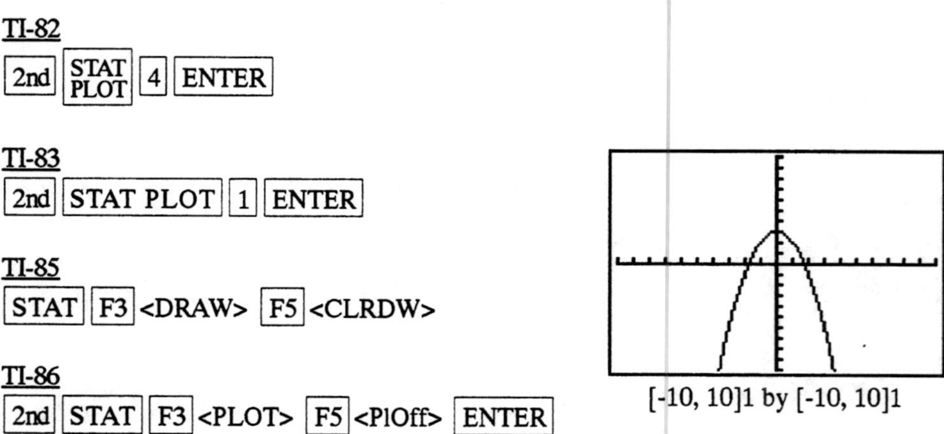

TI-82

2nd STAT PLOT 4 ENTER

TI-83

2nd STAT PLOT 1 ENTER

TI-85

STAT F3 <DRAW> F5 <CLRDW>

TI-86

2nd STAT F3 <PLOT> F5 <PlOff> ENTER

[-10, 10]1 by [-10, 10]1

Graph the function (see Appendix Section A-6, B-6, C-6 or D-6 of this manual).

EXAMPLE 3 ***TEXTBOOK SECTION 1-1***

Graph $y = \sqrt{4 - x}$ and determine the domain by examining the graph.

Solution: Store the function in the function list in the calculator as: √ (4 - X). Set the viewing window variables to [-5, 5]1 by [-5, 5]1 (see Appendix Section A-7, B-7, C-7 or D-7 Example 2 of this manual). Graph the function.

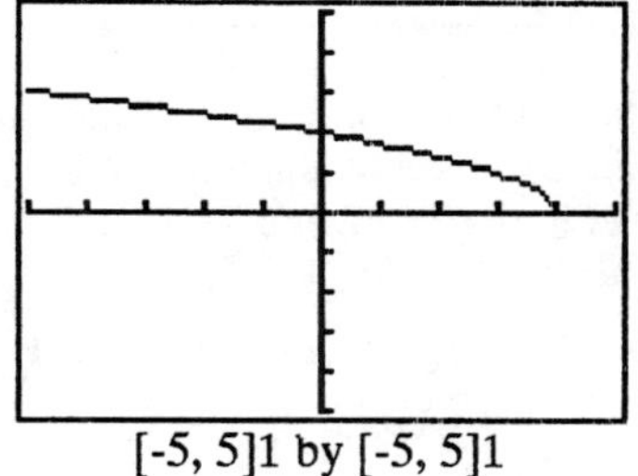
[-5, 5]1 by [-5, 5]1

Use the built-in function root or zero to find the x intercept (or root) of this function. (See Appendix Section A-7, B-7, C-7, or D-7 of this manual.)

The x intercept is 4. Hence the domain is $x \leq 4$.

EXAMPLE 4 ***TEXTBOOK SECTION 1-1***

(A) Fill in the chart below for $f(x) = 2x^2+3$.

(B) How does the quantity $\frac{f(2+h) - f(2)}{h}$ change as h assumes values .2, .1, .01, and .001.

(C) Make a conjecture as to what happens to this quantity when h gets closer and closer to zero.

h	.2	.1	.01	.001	→ 0
$\frac{f(2+h) - f(2)}{h}$					

Solution (A): $f(2+h) = 2(2+h)^2 + 3$ and $f(2) = 11$. Store $2(2+h)^2 + 3$ as 2 (2 + [ALPHA] H) ^ 2 + 3 as Y1 and 11 as Y2 in the function list on your calculator. Store (Y1-Y2)÷H as Y3 by entering:

TI-82

[(] [2nd] [Y-VARS] [1] [1] <Y1> [-] [2nd] [Y-VARS] [2] [2] <Y2> [)] [÷] [ALPHA] [X]

TI-83

[(] [VARS] [▶] [1] [1] <Y1> [-] [VARS] [▶] [1] [2] <Y2> [)] [÷] [ALPHA] [X]

TI-85 & TI-86

[(] [2nd] [ALPHA] [Y] [1] [-] [2nd] [ALPHA] [Y] [2] [)] [÷] [ALPHA] [H]

Store .2 as H and evaluate Y3 by recalling it (see Appendix Section A-4, B-4, C-4 or D-4 of this manual). Repeat by storing each of the other values of h and evaluating Y3. The results are shown below.

h	.2	.1	.01	.001	→	0
$\frac{f(2+h) - f(2)}{h}$	8.4	8.2	8.02	8.002	→	8

Note that the list feature of the calculator could be used by storing {.2, .1, .01, .001} as a list and then evaluating the function for this list. To do this you must use X in place of H in the TI-82 and TI-83 calculators and store the list as L1. To evaluate the function use y3(L1). The list can be stored as H on the TI-85 and TI-86 calculators. To evaluate y3 enter [2nd] [ALPHA] [Y] [3].

Solution (B) & (C): The quantity gets closer and closer to 8 as h gets closer and closer to 0.

EXAMPLE 5 ***TEXTBOOK SECTION 1-1***

A price-demand function that models a set of data is $p(x) = 94.8 - 5x$.

(A) Evaluate the price-demand function at $x = 2$, 5, 8, and 12 million.

(B) Find the error between the result of using the function to calculate the price and the actual data shown below.

x	p
2	87
5	68
8	53
12	37

(C) Plot the points and graph the price-demand function that models the data.

Solution (A): Store the function $p(x) = 94.8 - 5x$ as Y1 as 94.8 - 5 X. Store 2 as X. Recall the function Y1 and evaluate. Repeat for the other values of x. The list feature of the calculator could also have been used. (See Appendix Section A-4, B-4, C-4, or D-4 of this manual and Example 2 above.) The result are shown in the second column of the table to the right.

x	$p(x)$	p	*Error* $p(x)-p$
2	84.8	87	-2.2
5	69.8	68	1.8
8	54.8	53	1.8
12	34.8	37	-2.2

Solution (B): Calculate $p(x)$-p to find the error. The results are shown in the table above.

Solution (C): Set the viewing window variables at [0, 15]1 by [0, 100]10. (See Appendix Section A-8, B-8, C-8, or D-8 of this manual on how to choose and set the viewing window variables.) Plot the points (see Example 2 above). Store the price-demand function as Y1 as 94.8 - 5 X and graph.

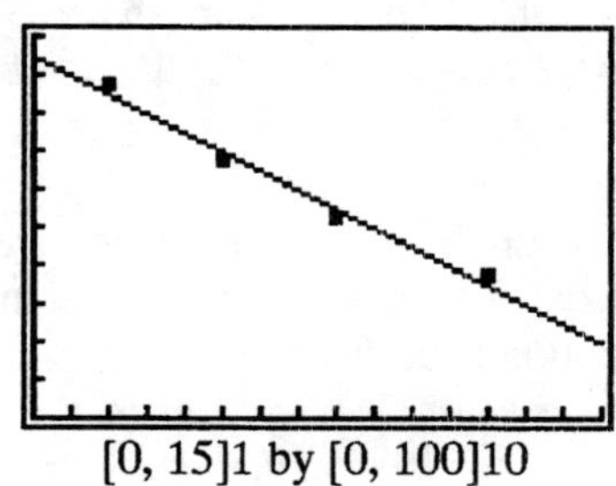

[0, 15]1 by [0, 100]10

EXAMPLE 6 ***TEXTBOOK SECTION 1-2***

Graph $f(x) = x^2$, $f(x) = x^2+4$, and $f(x) = x^2 - 7$ on the same coordinate axes. Compare the graphs and describe the effect of adding or subtracting a constant from $f(x)=x^2$.

Solution: Store the functions as three separate functions in the calculator as: X^2, X^2+4, and X^2-7. Graph using [-15, 15]1 by [-20, 20]1. We see that the second graph has the same shape but is shifted vertically upward 4 units from $f(x) = x^2$. The third graph has the same shape but is shifted vertically downward 7 units from $f(x) = x^2$.

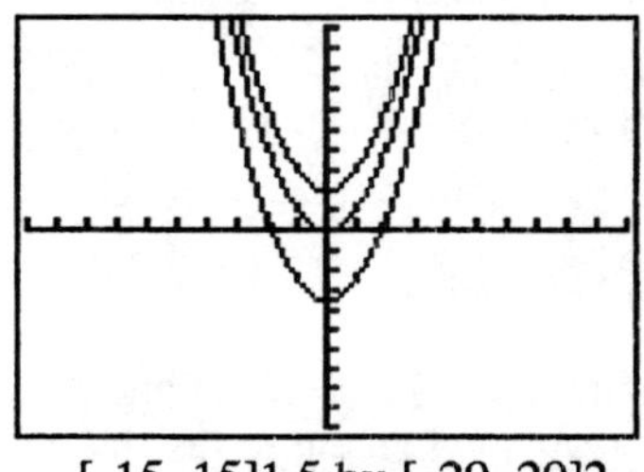

[-15, 15]1.5 by [-20, 20]2

EXAMPLE 7 ***TEXTBOOK SECTION 1-2***

(A) Find the value(s) of x for which $f(x) = |x|-5$ is positive.

(B) Discuss how the graph of this function differs from $y = |x|$.

Solution (A): Store $f(x)$ as Y1 as ABS (X) - 5 where ABS is the built-in absolute value function in your calculator (see Appendix Section A-2, B-2, C-2 or D-2 Example 1 of this manual). Set the viewing window variables to [-10, 10]1 by [-10, 10]1. Graph the function.

Zoom and trace, or use the root finder capability of your calculator, to find the x intercepts. (See Appendix Section A-7, B-7, C-7 or D-7 of this manual.) These roots are -5 and 5. The x intercepts are the cut-off values for the set of values for which $f(x)$ is positive.

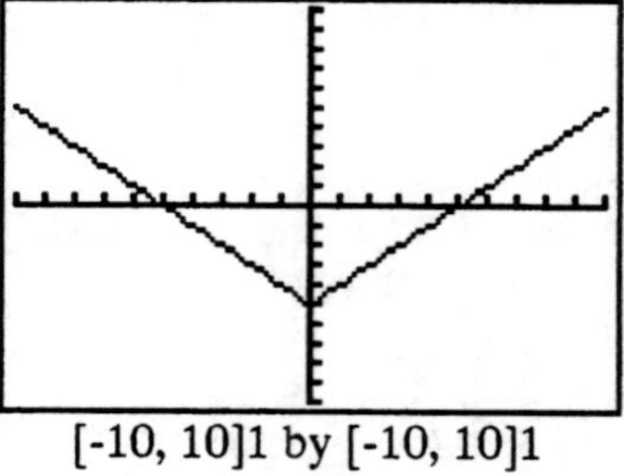

[-10, 10]1 by [-10, 10]1

Examine the graph to see where the graph is above the y axis. This is where the function is positive. The function is positive for x values greater than 5 and for x values less than -5. Hence, the solution is the set of all real numbers such that $x < -5$ or $x > 5$.

Solution (B): The graph of $f(x)$ differs from $y=|x|$ in that the graph of $f(x)$ is shifted 5 units downward from that of $y = |x|$.

EXAMPLE 8 ***TEXTBOOK SECTION 1-2***

Graph $f(x) = \begin{cases} 2x & x \le 1 \\ x^2 & 1 < x \le 2 \\ 1 - x & 2 < x \end{cases}$.

Solution: Store the function in the calculator as three separate pieces using an inequality statement to restrict the domain for each piece. The function will not be graphed for values of x where the function is undefined. Hence, we will divide by the inequality statement which causes the function to be only graphed for x values where it is defined. (See Appendix Section A-6, B-6, C-6, and D-6, of this manual.)

Y1: (2X)/(X≤1) [(] [2] [X] [)] [÷] [(] [X] [2nd] [TEST] [≤] [1] [)]

Y2: (X^2)/(1<X)/(X≤2) [(] [X] [^] [2] [)] [÷] [(] [1] [2nd] [TEST] [<] [X] [)] [÷] [(] [X] [2nd] [TEST] [≤] [X] [)]

Y1: (1-X^2)/(2<X) [(] [1] [-] [X] [)] [÷] [(] [2] [2nd] [TEST] [<] [X] [)]

Graph the function using the standard viewing window.

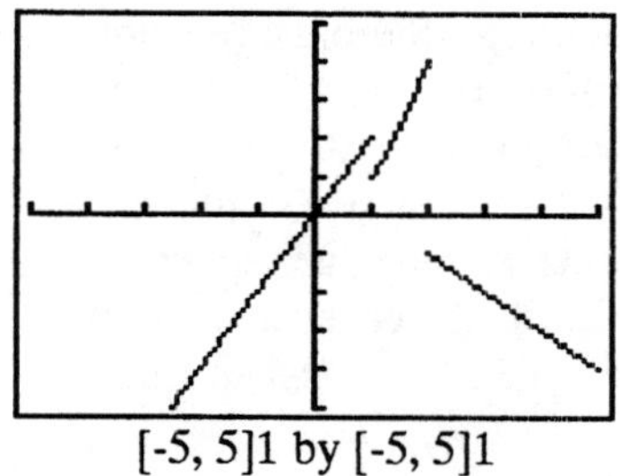

[-5, 5]1 by [-5, 5]1

EXAMPLE 9 ***TEXTBOOK SECTION 1-2 & 1-3***

Use graphing techniques to explore whether or not the two lines $3.15x + 2.79y = 5.38$ and $3.05x + 2.70y = 5.30$ are parallel.

Solution: Graph both lines on the same set of coordinate axes using [-10, 10]1 by [-10, 10]1. At first glance there appears to be only one line. However, use either the zoom box or zoom in features of the calculator, or set the graph screen at [.5, 1.5]0 by [.5, 3]0 to see that there are, in fact, two lines. The lines now appear to be parallel. However, their slopes vary slightly and hence are actually two intersecting lines. (The point of intersection is $(58, -63.\overline{5})$.

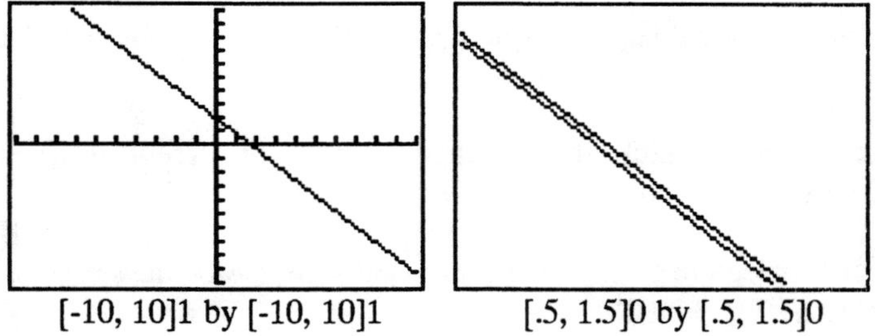

[-10, 10]1 by [-10, 10]1 [.5, 1.5]0 by [.5, 1.5]0

EXAMPLE 10 ***TEXTBOOK SECTION 1-2 & 1-3***

(A) Using a graphing utility, simultaneously graph $y = x + b$ for b = -5, -3, 0, 3, and 5 in a standard viewing window. Discuss the significance of the constant term.

(B) Using a graphing utility, graph $y = mx - 1$ for m = -2, -1, 0, 1, and 2 simultaneously in a standard viewing window. Discuss the significance of the slope.

Solution (A): Store the functions in the function list. Use [-10, 10]1 by [-10, 10]1 as the viewing window variables (see Appendix Section A-6, B-6, C-6 or D-6 of this manual). Graph. By examining the graph we see that the lines appear parallel. The slope of the line is 1 (the coefficient on x). The lines have y intercept equal to the constant term.

[-10, 10]1 by [-10, 10]1

Solution (B): Store the functions in the function list. Use [-10, 10]1 by [-10, 10]1 as the viewing window variables (see Appendix Section A-6, B-6, C-6 or D-6 of this manual). Graph. By examining the graph we see that the lines appear to have the same y intercept of -1 but have different slopes. The coefficient of x is the slope of the line.

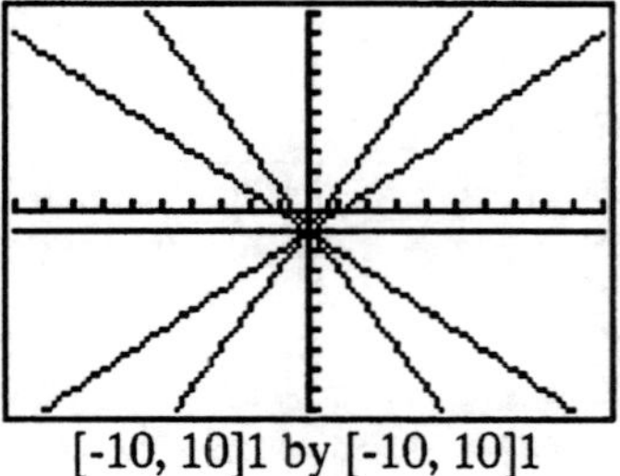

[-10, 10]1 by [-10, 10]1

EXAMPLE 11 ***TEXTBOOK SECTION 1-2 & 1-4***

(A) Graph $f(x) = -x^2 + 5x + 3$.

(B) Conjecture what you think will happen to the graph if the 3 is replaced by 6 in the function.

(C) Conjecture what you think will happen to the x and y intercepts if the 3 in $f(x)$ is replaced by a 6.

(D) Check your conjecture by graphing $g(x) = -x^2 + 5x + 6$ on the same set of coordinate axes.

Solution (A): Store $f(x)$ as Y1 in the calculator as - X ^ 2 + 5 X + 3. Set the viewing window variables at [-10, 10]1 by [-10, 10]1. Graph.

Solution (B): Let the new function be named $g(x) = -x^2 + 5x + 6$. A good conjecture is that the graph of $g(x)$ will be shifted upward 3 units from $f(x)$.

Solution (C): A good conjecture is that the least x intercept for $g(x)$ will be less than the least x intercept for $f(x)$ and that the greatest x intercept for $g(x)$ will be greater than the greatest x intercept for $f(x)$.

Furthermore, the y intercept of $g(x)$ will be greater than the y intercept for $f(x)$.

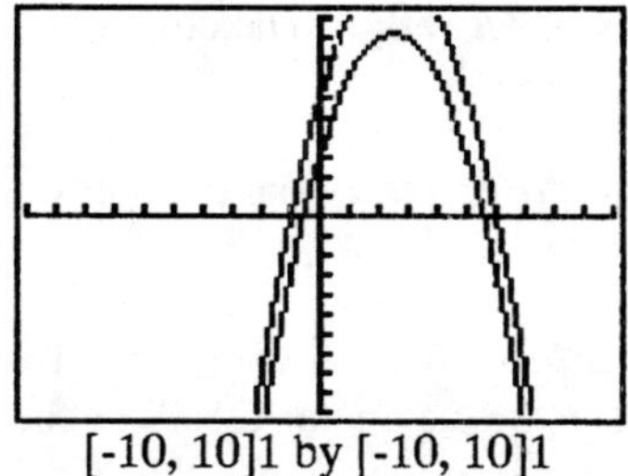

[-10, 10]1 by [-10, 10]1

Solution (D): The graph of $f(x)$ and $g(x)$ are shown to the right. See Appendix Section A-7 & A-8, B-7 & B-8, C-7 & C-8, or D-7 & D-8.

EXAMPLE 12 — *TEXTBOOK SECTION 1-2 & 1-4*

Consider the functions $f(x) = 2(x - 2)^2 + 3$ and $g(x) = 2(x - 4)^2 - 2$. Discuss the relationship of the graph of $f(x)$ to that of $g(x)$.

Solution: The graph of $g(x)$ will be two units to the left and 5 units downward from $f(x)$. The graphs are shown to the right.

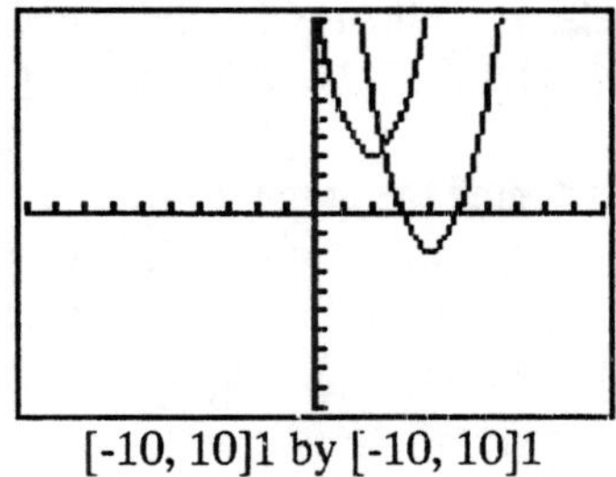

[-10, 10]1 by [-10, 10]1

EXAMPLE 13 — *TEXTBOOK SECTION 1-4*

The market research department of a company recommended to management that the company manufacture and market a promising new product. After extensive surveys, the research department backed up the recommendation with the price-demand equation $x = 4600 - 29p$ where x is the number of units that retailers are likely to buy per month at $\$p$ per unit for $0 \le p \le 150$, and where \$71,000 is the fixed cost and \$60 is the variable cost per unit.

(A) Find the cost function in terms of the price p.

(B) Find the revenue function in terms of the price p.

(C) Find the price p at which the company will break even. This price is where the cost is equal to the revenue.

Solution (A): The cost function is $C = 71000 + 60x = 71000 + 60(4600 - 29p)$.

Solution (B): The revenue equation is $R = xp = (4600 - 29p)p$.

Solution (C): We wish to find the price p at which the cost equals the revenue. Since the calculator uses X as the variable, change p to X and store the cost equation and the revenue equation in the calculator as separate functions as: 71000 + 60 (4600- 29 X) and (4600- 29 X) X .

The following chart will help in determining the viewing window variables.

x	Cost	Revenue
0	341000	0
50	254000	152500
100	167000	160000
150	80000	22500

A first choice for the viewing window variables could be [0, 150]10 by [100000, 200000]10000. The graph is shown to the right. (See Appendix Section A-8, B-8, C-8 or D-8 of this manual.)

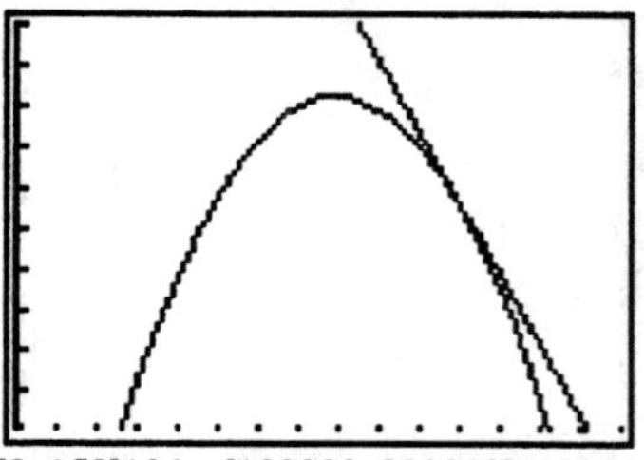
[0, 150]10 by [100000, 200000]100000

Examining the graph we see there appears to be one point of intersection. This is where the cost function and the revenue function have the same value. We want to find the X coordinate of this point.

Use the zoom feature of the calculator or change the viewing window variables to [104, 114]0 by [150000, 160000]0 to see that the graphs really do not intersect. This means that the revenue will never equal the cost. Hence, there is no break-even point. Since the revenue will always be less than the cost, this would not be a good product to produce based on this analysis.

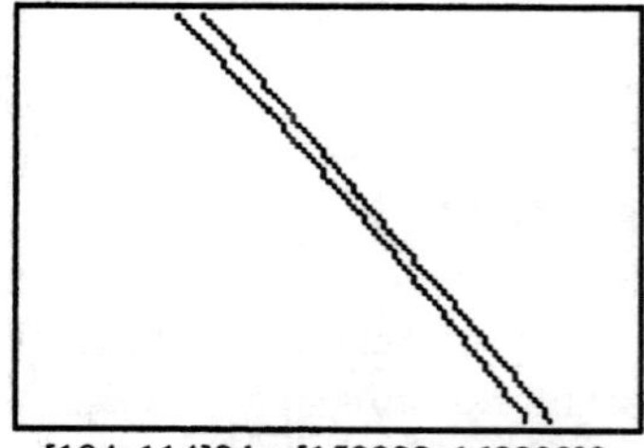
[104, 114]0 by [150000, 160000]0

This can be verified by attempting to solve the equation $71000 + 60(4600 - 29p) = (4600 - 29p)p$. This can be done algebraically by hand or using the built-in features (see Appendix Section A-9, B-9, C-9 or D-9 of this manual).

Exercise Set 1

1. Graph $f(x) = \sqrt{x^2 - 1}$. Use the graph to approximate the domain.

2. Graph $f(x) = \dfrac{\sqrt{x + 4}}{x - 1}$. Use the graph to approximate the domain.

3. Evaluate the $Q(x) = x^2 - 5x + 1$ for x = 2.5, 2.1, 2.01, and 2.001. Examine the results to determine what value $Q(x)$ gets close to as x gets close to 2.

4. Repeat Problem 3 above for x = 1, 1.5, 1.9, 1.99, and 1.999. Examine the results to determine what value $Q(x)$ gets close to when x gets close to 2.

5. Let $f(x) = x^2 - 2.3x + 3$. How does the quantity $\dfrac{f(1.5+h) - f(1.5)}{h}$ change as h assumes values .2, .1, .01, and .001? Fill in the chart and make a conjecture as to what happens to this quantity when h gets closer and closer to zero.

h	.2	.1	.01	.001	→	0
$\dfrac{f(1.5+h) - f(1.5)}{h}$						

6. Let $f(x) = 3x^2 - 1.3x + 3$. How does the quantity $\dfrac{f(.8+h) - f(.8)}{h}$ change as h assumes values .2, .1, .01, and .001. Fill in the chart and make a conjecture as to what happens to this quantity when h gets closer and closer to zero.

h	.2	.1	.01	.001	→	0
$\dfrac{f(.8+h) - f(.8)}{h}$						

7. Repeat Example 5 above for $p(x) = 1190 - 36x$ and the values given to the right. Find the error between the price-demand function $p(x)$ that models the data and the price p for each value of x.

x	p
2	87
5	68
8	53
12	37

8. Graph the following functions and explain how each is related to the graph of one of the six basic functions.

	Transformed Function	Basic Function
(A)	$g(x) = -\lvert x + 3\rvert$	$f(x) = \lvert x\rvert$
(B)	$g(x) = (x - 4)^2 - 3$	$f(x) = x^2$
(C)	$g(x) = 7 - \sqrt{x}$	$f(x) = \sqrt{x}$
(D)	$g(x) = -0.4x^2$	$f(x) = x^2$

9. (A) Graph $f(x) = 1.2x - 4.2$ in a standard viewing window.

   (B) Find the x and y intercepts to one decimal place using trace and zoom or an appropriate built-in function of your calculator.

   (C) Find the x and y intercepts algebraically to one decimal place.

   (D) Using the results of parts (A)-(C), find the solution set for the linear inequality $1.2x - 4.2 > 0$.

   (E) Conjecture, then check, what the solution to Part (D) would be if the function was $g(x) = 2.2x - 4.2$.

10. Repeat Problem 9 using $f(x) = 1.2x - 3.2$.

11. Graph the following equations simultaneously in a standard viewing window:

    $3x + 2y = 6 \qquad 3x + 2y = 3 \qquad 3x + 2y = -6 \qquad 3x + 2y = -3.$

    Conjecture, then check, if the graph of $3x + 2y = 0$ is in this family of lines.

12. Graph the following two equations simultaneously in a standard viewing window:

    $3x + 4y = 12 \qquad 4x - 3y = 12.$

    Conjecture, then check, how the graphs of $3x + 4y = 0$ and $4x - 3y = 0$ are related.

13. Graph the piecewise-defined function and find any points of discontinuity:

$$\text{Graph } f(x) = \begin{cases} 2x - 1 & x \le 0 \\ x^2 + 2 & 0 < x \le 2 \\ 4 + x & 2 < x \end{cases} .$$

14. Graph the piecewise-defined function and find any points of discontinuity:

$$\text{Graph } f(x) = \begin{cases} -3x & x \le -1 \\ \sqrt{x^2+8} & -1 < x \le 1 \\ 4 + x & 1 < x \end{cases}$$

15. Use graphing techniques to explore whether or not the two lines $5.23x - 8.92y = 6.34$ and $5.21x - 8.89y = 6.30$ are parallel.

16. Use graphing techniques to explore whether or not the two lines $5.23x - 8.92y = 6.34$ and $y = .60x + 6.38$ are parallel.

17. Graph $f(x) = x^2$, $f(x) = 3x^2$, and $f(x) = -1.5x^2$ on the same coordinate axes. Compare the graphs and describe the effect of multiplying $f(x)=x^2$ by a constant.

18. Graph $f(x) = x^2$, $f(x) = (x-1)^2$, and $f(x) = (x+2)^2$ on the same coordinate axes. Compare the graphs and describe the effect of adding/subtracting a constant from the x term of $f(x)=x^2$.

NOTES

NOTES

Chapter 2

Additional Elementary Functions

This chapter contains examples using the calculator to illustrate:

- Exploring asymptotic behavior of a function
- Approximating zeros or roots of a polynomial
- Finding regression models
- Determining the domain of a function
- Exploring discontinuities of a function
- Exploring vertical asymptotes of a function
- Exploring horizontal asymptotes of a function
- Evaluating exponential and logarithmic functions
- Graphing exponential and logarithmic functions
- Solving problems involving growth or decay
- Graphing inverse functions
- Using the change of base formula to evaluate logarithms

EXAMPLE 1 ***TEXTBOOK SECTION 2-1***

How does the graph of $f(x) = x^4 - 3x^2 + x + 2$ compare to the graph of $y = x^4$ as we "zoom out?"

Solution:

Graph both functions using a viewing window of [-10, 10]1 by [-10, 10]1. [-10, 10]1 by [-40, 40]4, and [--10, 10]1 by [-100, 100]10. We see that $f(x)$ gets close to y for larger values of x.

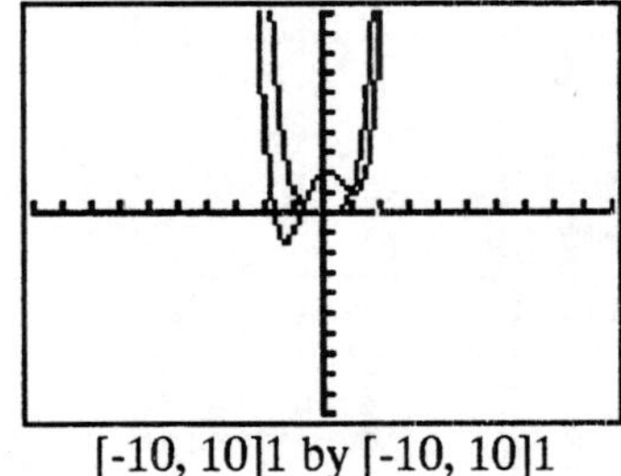

[-10, 10]1 by [-10, 10]1

Repeat using [-10, 10]1 by [-40, 40]4. and [-10, 10]1 by [-100, 100]10. We see that $f(x)$ gets close to y for larger values of x.

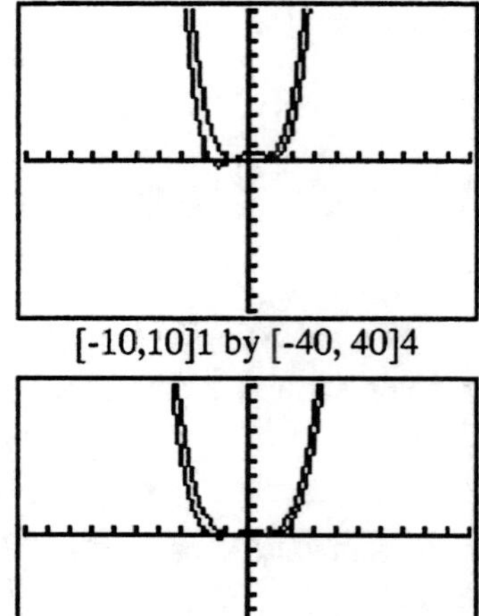

[-10,10]1 by [-40, 40]4

Repeat using [-10, 10]1 by [-100, 100]10.

We see that $f(x)$ gets close to y as x gets larger. In the graphs shown at the right, we needed to increase the range (values of y) to see how the graphs behaved as x increased since the values of the function were so large for smaller values of x.

[-10, 10]1 by [-100, 100]1

EXAMPLE 2 ***TEXTBOOK SECTION 2-1***

Approximate, to two decimal places, the zeros of $f(x) = 2x^4 - 5x^3 - 4x^2 + 3x + 6$.

Solution: Graph the function by storing it as Y1 in the calculator. Set the viewing window variables at [-10, 10]1 by [-10, 10]1. Use the root finder feature of the calculator, or store the function as Y1, store 0 as Y2, and use the intersect feature of the calculator. (See Appendix Section A-7, B-7, C-7 or D-7 of this manual.)

The x intercepts are 1.15 and 2.89.

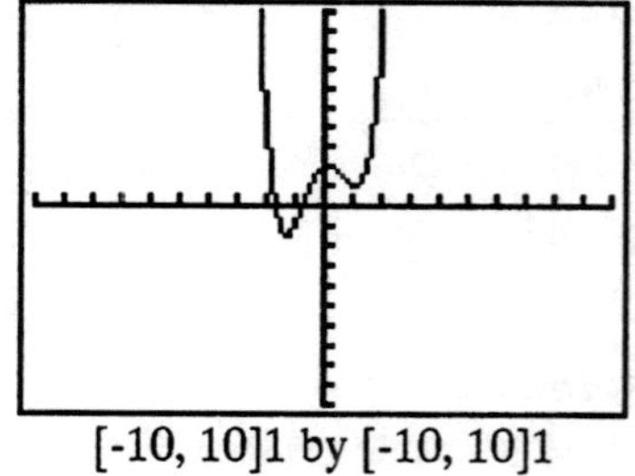

[-10, 10]1 by [-10, 10]1

EXAMPLE 3 ***TEXTBOOK SECTION 2-1***

Given the following set of data:

x	10	14	18	22	26	30	34	38	44
y	5	12	26	56	96	152	226	326	536

(A) Find a cubic polynomial regression model.

(B) Graph the data points and the regression model on the same set of coordinate axes.

(C) Predict the value of y for $x = 50$ using this regression model.

Solution (A): Enter the data into the statistical lists in the calculator. (See Appendix Section A-16, B-16, C-16 or D-16 Example 2 of this manual.) Calculate the cubic regression model. The result is $y = .01x^3 - .21x^2 + 3.23x - 17.54$ where the coefficients have been rounded to two decimal places.

Solution (B): Store the regression equation in the Y= list. Set the viewing window variables at [0, 50]5 by [0, 600]100. Graph the function and the points at the same time. (See Appendix Section A-16, B-16, C-16, or D-16 Example 2(C) of this manual.)

Solution (C): Store 50 as x in the calculator. Recall the function from the Y= list or RegEq from the VARS list and evaluate. The result is 816.38 rounded to two decimal places.

EXAMPLE 4 ***TEXTBOOK SECTION 2-1***

Consider the function $f(x) = \frac{x^2 - 4}{x - 2}$.

(A) Graph using viewing window variables of [-4.7, 4.7]1 by [-1, 5.2]1 on the TI-82 and TI-83, or [-6.3, 6.3]1 by [-1, 5.2]1 on the TI-85 and TI-86. Note that by using these viewing window variables the coordinates for x and y at the bottom of the viewing window are in steps of .1 when you use the arrow keys to find screen locations on the calculator. This is because on the TI-82 and TI-83 the screen is 95 pixels across by 63 pixels down; and on the TI-85 and TI-86 the screen is 127 pixels across by 63 pixels down.

(B) Find the domain.

Solution (A): Enter the function in the function list as (X^2-4)÷(X-2). Set the viewing window variables and graph. Note that there is a hole in the graph at $x = 2$. Use the trace feature of the calculator to see this.

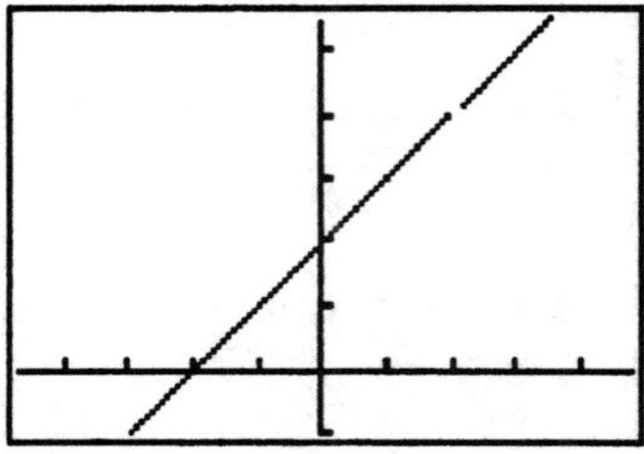

Solution (B): When you use the trace feature of the calculator you see that when the x coordinate is 2 there is no coordinate for y at the bottom of the screen. This indicates that 2 is not in the domain of the function.

Another way to see this is to return to the home screen, store 2 as X, and evaluate the function. An error message occurs indicating that the function cannot be evaluated for this value of x. This can also be seen be examining the denominator of $f(x)$. The denominator is equal to zero when $x = 2$. Hence the domain of $f(x)$ is $D = \{x \mid x \neq 2\}$.

EXAMPLE 5 ***TEXTBOOK SECTION 2-1***

Consider the function $f(x) = \dfrac{3x^2 + 5x - 8}{x^2 - 1}$.

(A) Graph using [-4.7, 4.7]1 by [-12.4, 12.4]1 on the TI-82 and TI-83, or [-6.3, 6.3]1 by [-12.4, 12.4]1 on the TI-85 and TI-86. Note that by using this range the coordinates for x at the bottom of the viewing window are in steps of .1 and y is in steps of .4 when you use the arrow keys to find screen locations (not using the trace feature) on the calculator This is because the screen on the TI-82 and TI-83 it is 95 pixels across by 63 pixels down; and on the TI-85 and TI-86 the screen is 127 pixels across by 63 pixels down.

(B) Find the domain.

Solution (A): Enter the function in the function list as (3X^2+5X-8)÷(X^2−1). Set the viewing window variables and graph the function. Note that there is unusual behavior when $x = -1$ (a vertical asymptote) and $x = 1$ (a hole in the graph).

Solution (B): Use trace to see that there is no y coordinate when the x coordinate is 1 or -1. This indicates that -1 and 1 are not in the domain of the function.

Another way to see this is to return to the home screen, store 1 as X, and evaluate the function. An error message occurs indicating that the function cannot be evaluated for this value of x. This can also be seen be examining the denominator of $f(x)$. The denominator is equal to zero when $x = 1$. Similarly for $x = -1$.

Hence the domain of $f(x)$ is $D = \{x \mid x \neq 1, x \neq -1\}$.

EXAMPLE 6 ***TEXTBOOK SECTION 2-1***

(A) Investigate the behavior of $f(x) = \dfrac{3x^2 + 5x - 8}{x^2 - 1}$ when x is close to 1 and -1.

(B) Write summary statements to Part (A) symbolically by filling in the blanks in the notation $f(x) \to$ ___ as $x \to 1$ and $f(x) \to$ ___ as $x \to -1$.

(C) Write the equations of any vertical asymptotes.

(D) Investigate the graph of f as x increases or decreases without bound.

Solution: We shall conduct our investigation by constructing a table for $f(x)$ for x values near but not equal to the given value and by examining the graph of the function using the trace and zoom features of the calculator.

Solution (A): Store the function in the calculator as (3 X ^ 2 + 5 X - 8) ÷ (X ^ 2 - 1). (See Appendix Section A-4, B-4, C-4 or D-4 of this manual.) Store .9 as X and evaluate the expression. The result is 5.632, rounded to the nearest thousandth. Repeat for the other values of x. [NOTE: The list feature could be used in evaluating this function.]

x	.900	.990	.999	→ 1 ←	1.001	1.010	1.100
$f(x)$	5.632	5.513	5.501→ ?		?←5.499	5.488	5.381

The table shows that the function is approaching 5.5 as x approaches 1 from both the right and the left. Hence, $\lim_{x \to 1} f(x) = 5.5$. However, as we saw in Example 3 above, when 1 is stored for x and the function is evaluated we get an error message. The denominator has value 0 when $x = 1$ and the calculator is attempting to divide by zero. This indicates that the function is not defined for $x = 1$. $x = 1$ is not in the domain of the function.

Set the viewing window variables as in Example 3 above graph the function that is stored. Use the trace feature of the calculator and move the cursor along the curve for values of x near 1. Note the y values are near 5.5. Now zoom in on the point of the graph where x is close to 1 or change the viewing window variables to [.90, 1.10].01 by [5.3, 5.7].1 and trace again.

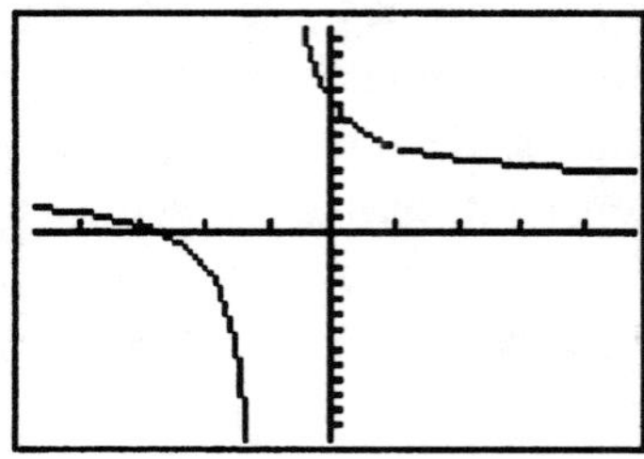

We do not see a hole in the graph displayed on the calculator screen since the x values on the screen do not equal 1 exactly. The graph will show a hole at $x = 1$ if the viewing window variables are set as in Example 3 above.

Repeat the investigation for values of x near -1. Make sure the function is still stored in your calculator.

x	-1.100	-1.010	-1.001	→ -1 ←	-.999	-.990	-.900
$f(x)$	-47	-497	-4997→ ?		?←5003	503	53

The table shows that the function values are getting greater and greater negatively for values of x close to but a little less than -1. The function values are getting greater and greater positively for values of x close to but a little greater than -1.

Set the viewing window variables to [-10, 10]1 by [-10, 10]1 and graph the function that is stored. Use the trace feature of the calculator and move the cursor along the curve for values of x near -1. We see that the y values get very large in the negative direction as x gets close to but is less than -1. Then as the cursor is moved from a little less than -1 to x values a little greater than -1 the y values jump to very large positive numbers. However, as we saw in Example 2 above, when -1 is stored for x and the function is evaluated we get an error message. The denominator has value 0 when $x = -1$ and the calculator is attempting to divide by zero. This indicates that the function is not defined for $x = -1$. $x = -1$ is not in the domain of the function.

Solution (B): Symbolically we write $f(x) \to 4$ as $x \to 1$ and $f(x) \to -\infty$ as $x \to -1^-$ $f(x) \to \infty$ as $x \to -1^+$.

Solution (C): The line $x = -1$ is a vertical asymptote.

Solution (D): To investigate the function as x increases without bound let us make a table. Store the function as Y1 in the calculator. Store the value of 10 as X and evaluate the function. Repeat for each value of X or use the list feature of your calculator. (See Appendix Section A-4, B-4, C-4 or D-4 of this manual.)

x	10	100	1,000	10,000	→	∞
$f(x)$	3.4545	3.0495	3.0050	3.0005	→	3

The graph of $y = f(x)$ approaches the line $y=3$ from above as x increases without bound. This can also be seen by using the trace feature of the calculator to exam the graph. Use the right arrow key to move the cursor far to the right. Watch the values of the y coordinate as you do this.

We repeat the process to investigate the function as x decreases without bound. We complete the table below.

x	-10	-100	-1,000	-10,000	→	$-\infty$
$f(x)$	2.4444	2.9495	2.9950	2.9995	→	3

The graph of $y = f(x)$ approaches the line $y = 3$ from below as x decreases without bound. This can also be seen by using trace on the calculator to exam the graph. Use the left arrow key to move the cursor far to the left. Watch the values of the y coordinate as you do this.

EXAMPLE 7 — *TEXTBOOK SECTION 2-1*

(A) Investigate the behavior of $f(x) = \frac{|x - 2|}{x - 2}$ as x is near 2.

(B) Does $f(x)$ have a discontinuity at $x = 2$?

Solution (A): We shall conduct our investigation by constructing a table for $f(x)$ for x values near but not equal to 2 and by examining the graph of the function for x near 2 using the trace and zoom features of the calculator.

Store the function in the calculator as ABS (X - 2) ÷ (X - 2) where ABS is the built-in absolute value function in your calculator. (See Appendix Section A-2 and A-4, B-2 and B-4, C-2 and C-4, or D-2 and D-4 of this manual.) Store each value of x in the calculator and evaluate the stored function or use the list feature of the calculator.

x	1.900	1.990	1.999	→ 2 ←	2.001	2.010	2.100
$f(x)$	-1	-1	-1 → ?	?←	1	1	1

We see that the function values are -1 for values of x less than 2. The function values are 1 for values of x greater than 2.

Set the calculator in dotted mode. Set the viewing window variables to [-10, 10]1 by [-10, 10]1 and graph. Use the trace feature of the calculator and move the cursor near $x = 2$. Watch the y values as x moves from a little less than 2 to a little greater than 2. The y value jumps from -1 to 1.

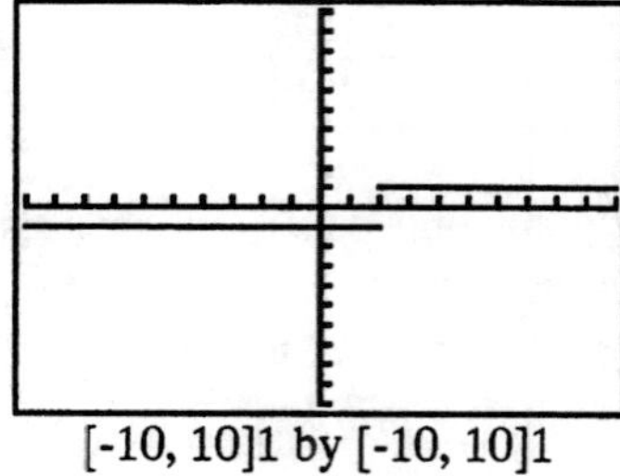

[-10, 10]1 by [-10, 10]1

Solution (B): Substitute 2 for x in the calculator and evaluate the function. We get an error message. This is because the calculator is trying to evaluate $\frac{0}{0}$. Hence we see that $f(x)$ is discontinuous at $x = 2$.

EXAMPLE 8 ***TEXTBOOK SECTION 2-2***

Graph $f(x) = 1.2^x$, $g(x) = 2.5^x$, $h(x) = -4^x$, and $k(x) = 4^x$ on the same set of coordinate axes. Describe the differences between the graphs.

Solution: Store the four functions in the calculator as 1.2^X, 2.5^X, -4^X, and 4^X. Note that -4^x is not the same function as $(-4)^x$. In fact, -4 cannot be the base of an exponential function.

Graph first using [-10, 10]1 by [-10, 10]1. It is hard to see the difference between the graphs. Change the viewing window variables to [-3, 3]1 by [-5, 5]1 to get the graph shown at the right.

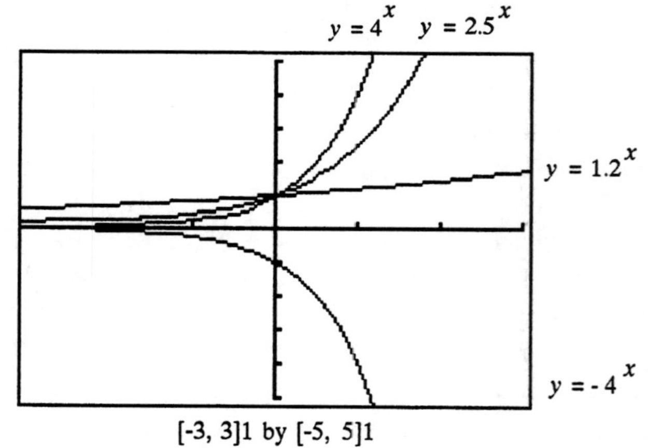

[-3, 3]1 by [-5, 5]1

The graph $f(x) = 1.2^x$ is lower than $g(x) = 2.5^x$ when $x > 0$, equal to $g(x) = 2.5^x$ when $x = 0$

and higher than $g(x) = 2.5^x$ when $x < 0$. A similar comparison can be made between $g(x) = 2.5^x$ and $k(x) = 4^x$. The graph of $h(x)$ is reflected about the x axis from $k(x)$.

EXAMPLE 9 ***TEXTBOOK SECTION 2-2***

A group of 400 parents, relatives, and friends are waiting anxiously at an airport for a student charter flight to return after a four-week trip to Europe. It is stormy and the plane is late. A particular parent thought he had heard that the plane's radio had gone out and related this news to some friends, who in turn passed it on to others, and so on. Sociologists have studied rumor propagation and have found that a rumor tends to spread according to: $N(t) = \dfrac{400}{1+399e^{-0.4t}}$ where t is measured in minutes.

(A) Graph this function.

(B) How many people will have heard the rumor after 5 minutes?

(C) Approximate the amount of time for 50% of the people to have heard the rumor.

(D) Approximate the amount of time for 75% of the people to have heard the rumor.

(E) Approximate the amount of time it will take for everyone to hear the rumor.

Solution: (A) Change t to X and store the function in the calculator as: 400 ÷ (1 + 399 **e**^ (⁻ .4 X)). (Recall e^ is the inverse ln function.)

We need to determine the viewing window variables. Evaluate the function for some values of X. Round numbers up since $N(x)$ represents the number of people having heard the rumor. For example, $N(10) = 48.146$ means that 49 people heard the rumor.

x	$N(x)$
0	1
10	49
20	353
30	400

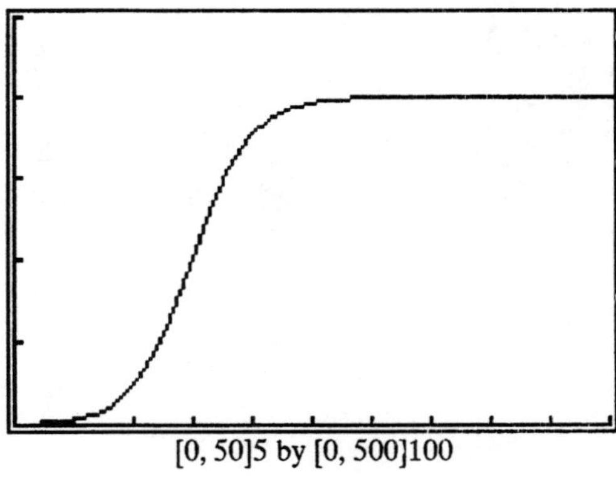

[0, 50]5 by [0, 500]100

From this table we see a good starting viewing window is [0, 50]5 by [0, 500]100.

(B) Evaluate the function for $x = 5$. The result is 8 people.

(C) Use the trace feature of the calculator and move the cursor to a position on the curve where the y value is close to 200 since 200 is 50% of 400. Zoom in on this value until accuracy is to one decimal place and round up. The value of x is approximately 15. It takes approximately 15 minutes for one-half of the people to have heard the rumor.

(D) Use the trace feature of the calculator and move the cursor to a position on the curve where the y value is close to 300 since 300 is 75% of 400. Zoom in on this value until accuracy is to one decimal place and round up. The value of x is approximately 18. It takes approximately 18 minutes for three-fourths of the people to have heard the rumor.

(E) Change the viewing window variables back to [0, 50]5 by [0, 500]100. Now use trace and watch the y variable as x increases. We see that as x gets larger and larger the y value gets closer and closer to 400. This makes sense since all people will hear a rumor eventually. We also see that 400 is reached when $x \approx 32$ which means that everyone will have heard the rumor in about 32 minutes.

EXAMPLE 10 — *TEXTBOOK SECTION 2-3*

(A) Graph $y = 3^x$, its inverse, and $y = x$ on the same set of coordinate axes. (B) Calculate the y coordinate for $y = 3^x$ when $x = 2$. Calculate the y coordinate for $y = \log_3 x$ when $x = 9$. Mark the points on the respective graphs. Discuss the relationship between these two pairs of coordinates.

Solution (A): First we need to write the inverse function of $y = 3^x$. We want to graph $y = \log_3 x$. The calculator can only calculate or graph functions having base 10 or base *e*. However, there is a formula that allows us to change from one base, say *b*, to another base, say *c*. This formula is: $\log_b x = \dfrac{\log_c x}{\log_c b}$. In our situation we have:

$\log_3 x = \dfrac{\log_{10} x}{\log_{10} 3}$. Hence we will store the three functions separately in the calculator as

3^X, X, and ([LOG] X) ÷ ([LOG] 3). Graph using [-6, 6]1 by [-4, 4]1. Note the y axis limits are 2/3 those of the x axis. This will make the scale marks about the same distance apart on both axes. The graph is shown at the right below.

Solution (B): For $y = 3^x$, $y = 9$ when $x = 2$. So the first coordinate pair is (2, 9).

For $\log_3 x$, $y = 2$ when $x = 9$. The second coordinate pair is (9, 2).

The coordinates are interchanged between the function and its inverse. The points are shown on the graph above. They are equidistant from the line $y = x$.

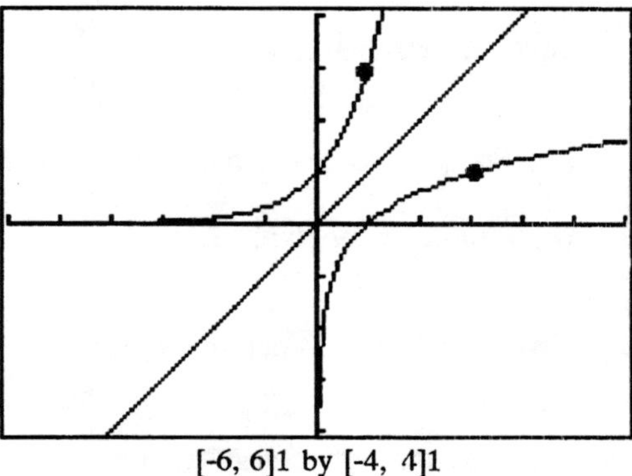

[-6, 6]1 by [-4, 4]1

Exercise Set 2

1. Approximate the zeros of $f(x) = x^5 - 12x^4 + 8x^3 + 11$ to two decimal places.

2. Approximate the zeros of $f(x) = x^5 + 14x^4 - 12x^3 - 13$ to two decimal places.

3. The following table shows the total national expenditures (in billion \$) and the percapita expenditures (in dollars) for selected years since 1960 (U. S. Census Bureau).

Date	1960	1965	1970	1975	1980	1985	1990
Total Expenditures (billion \$)	27.1	41.6	74.4	132.9	247.2	428.2	697.5
Per Capita Expenditures	143	204	346	592	1002	1666	2588

   (A) Let x represent the number ofyears since 1960 and find a cubic regression polynomial for the total national expenditures.

   (B) Use the regression polynomial found in Part (A) to estimate the total national expenditures (to the nearest hundred thousand) for 1995.

4. Use the data from Problem 3 to find a cubic regression polynomial for the per capita expenditures and estimate the percapita expenditures (to the nearest dollar) for 1995.

5. Given the rational function $f(x) = \dfrac{2x^2 - 7x + 5}{x^2 - 2x + 1}$.

   (A) Find any vertical and horizontal asymptotes for the graph.

   (B) Graph the function.

6. Given the rational function $f(x) = \dfrac{2x^2 - 3x - 5}{x^2 - 4x + 3}$.

   (A) Find any vertical and horizontal asymptotes for the graph.

   (B) Graph the function.

7. Graph $f(x) = 2.1^{x-1}$, $g(x) = 2.1^{x-2}$, $h(x) = 2.1^{x+1}$, and $j(x) = 2.1^{x}$ on the same graphing screen. Describe the differences between the graphs.

8. Graph $f(x) = 2.1^{x}$, $g(x) = 2.1^{3x}$, and $h(x) = 2.1^{-2x}$ on the same graphing screen. Describe the differences between the graphs.

9. Evaluate to five decimal places $\log_{1.3} 368.93$.

10. Evaluate to five decimal places $\log_{12.3} 368.93$.

11. Graph $y = 4^x$, its inverse, and $y = x$ on the same set of coordinate axes.

12. Graph $y = 1.5^x$, its inverse, and $y = x$ on the same set of coordinate axes.

NOTES

Chapter 3

Mathematics of Finance

This chapter contains examples using the calculator to illustrate:

- Calculation of simple interest, compound interest, future value of an ordinary annuity, and present value of an ordinary annuity
- Graphing of the future value with simple interest and compound interest
- Approximation of the interest rate in future value or amortization
- Evaluation of an algebraic expression to find future value and amortization schedules

EXAMPLE 1 ***TEXTBOOK SECTION 3-1***

$38,597 is invested in an account earning simple interest of 3.57%.

(A) How much interest will be earned if the money is left in the account for one year? Two years? Five years?

(B) Graph the future value as a function of time.

The replay key allows rapid calculation of expressions that are similar. This key will replay the last executed expression so you can change a quantity and recalculate. (Press [2nd] [ENTRY].)

Solution (A): Method 1 Each calculation can be done separately on the home screen without storing a formula. Enter 38597×.0357×1 in the calculator and evaluate. The result is $1,377.91. Use the replay key and change the 1 to a 2. Evaluate again. The result is $2,755.83. Use the replay key again and change the 2 to a 5. Evaluate again. The result is $6,889.56.

Method 2 Return to the home screen by pressing [2nd] [QUIT]. The future value of this loan after *t* years is A = 38597(1+.0357t). Change the *t* to an X and store the function as Y1 in the calculato as 38597 (1 + .0357 * X) as Y1 in the function list. Store the numbers 1, 2, and 5 in a list and evaluate the expression (see Appendix Section A-4, B-4, C-4 or D-4 of this manual).

Solution (B): The future value of this loan after t years is A = 38597(1+.0357t). Change the t to an X and store the function as Y1 in the calculator. An appropriate setting for the viewing window variables would be [0, 10]1 by [0, 100000]10000. The graph is shown at the right.

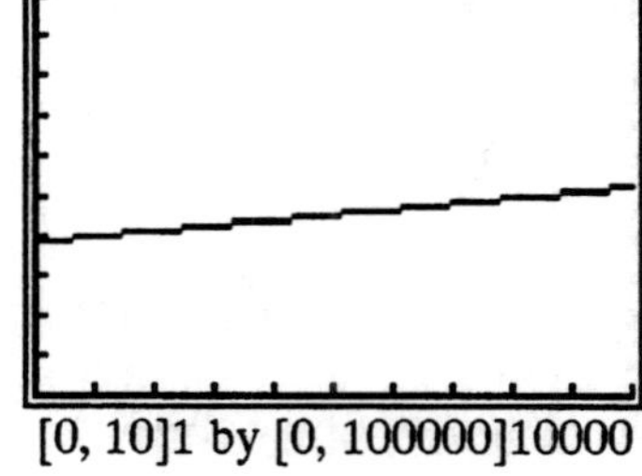
[0, 10]1 by [0, 100000]10000

Note that this graph implies interest is accumulating continuously when in fact it is calculated at the end of each time period.

EXAMPLE 2 ***TEXTBOOK SECTION 3-1 & 3-2***

$10000 is deposited in an account.

(A) Calculate the amount in the account at the end of 30 years if the account earns 8% simple interest.

(B) Calculate the amount in the account at the end of 30 years if the account earns 8% interest compounded yearly.

(C) Calculate the amount in the account at the end of 30 years if the account earns 8% interest compounded monthly.

(D) Calculate the amount in the account at the end of 30 years if the account earns 8% interest compounded daily.

(E) Graph the growth of the accounts for Part (B), (C) and (D) using viewing window variables of [0, 30]5 by [0, 111000]0. Compare the graphs. To see the difference between the graphs for monthly and daily compounding, set the viewing window variables at [29.9, 30].1 by [109000, 110400]0.

Solution: Set the number of decimal places to two. (See Appendix Section A-17, B-17, C-17 or D-17 of this manual.)

Solution (A): The formula for calculating future value for simple interest is $A = P(1+rt)$ where A =future value, P=principal, r=yearly interest rate, and t=number of years. For this situation we have 10000(1+.08×30) = 34000. Enter this on the home screen and calculate by pressing [ENTER]. $34,000 is in the account after 30 years.

Solution (B): The formula for calculating the future value for compound interest is $A = P(1 + \frac{r}{n})^{nt}$ where A =future value, P=principal, r=yearly interest rate, n=number of compounding periods in one year, and t=number of years. For this situation we have $10000\left(1 + \frac{.08}{1}\right)^{1\times30}$ = 100626.57. Enter this on the home screen as 10000 (1 + .08 ÷ 1) ^ (1 * 30) and calculate by pressing [ENTER]. There is $100,626.57 in the account after 30 years.

Solution (B) on the TI-83 calculator:
The TI-83 calculator has financial functions that will calculate compound interest when certain variables are stored in the calculator. These are explained below for Solution (B).

[2nd] [FINANCE] [1] <TVM Solver...>	To get the FINANCE CALC menu and TVM Solver.
[30] [ENTER]	To set N=30, the number of years.
[8] [ENTER]	To set I%=8, the interest rate.
[(-)] [10000] [ENTER]	To set PV=-10000, as a cash outflow which is the present value deposited.
[0] [ENTER]	To set PMT=0 since there is no payment each period.
[ENTER]	Do not set FV since this is what we are looking for.
[1] [ENTER]	To set P/Y=1, the number of payment periods per year.
[1] [ENTER]	To set C/Y=1, the number of compounding periods per year.
[ENTER]	So the calculation of interest will be done at the end of the compounding period.
[▲] [▲] [▲] [ALPHA] [SOLVE]	To place the cursor on FV and solve for the future value.

Solution (C): Using the formula in Part (B) we have $10000\left(1 + \frac{.08}{12}\right)^{12\times30}$ = 109357.30. Use the above calculator instructions but set P/Y=1 and C/Y=12. Solving for PV we find that there is $109,357.30 in the account after 30 years.

Solution (D): Using the formula in Part (B) we have $10000\left(1 + \frac{.08}{365}\right)^{365\times30}$ = 110202.78. Use the above calculator instructions but set P/Y=1 and C/Y=365. Solving for PV we find that there is $110,202.78 in the account after 30 years.

Solutions (C) and (D) on the TI-83 calculator: These can be found by changing the number of compounding periods per year, P/Y, to the appropriate value: P/Y=1 and C/Y=12 for Solution (C) and P/Y=1 and C/Y=365 for Solution (D).

Solution (E): Store the following in the function list in the calculator as:

Y1	10000 (1 + .08 ÷ 1) ^ (1 X)
Y2	10000 (1 + .08 ÷ 12) ^ (12 X)
Y3	10000 (1 + .08 ÷ 365) ^ (365 X)

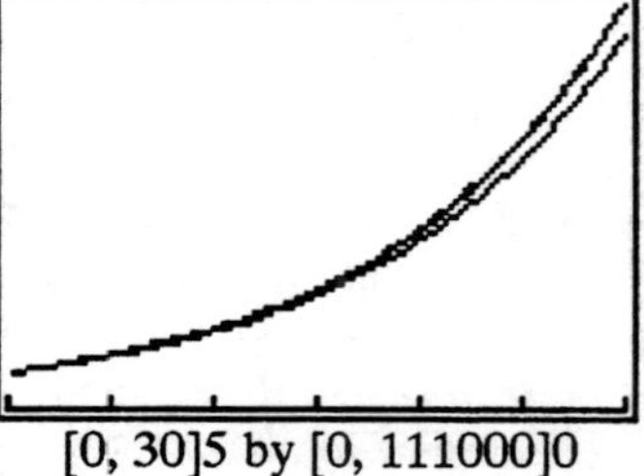
[0, 30]5 by [0, 111000]0

Set the viewing window variables to [0, 30]5 by [0, 111000]0 and graph. Watch very closely as Y2 and Y3 are graphed. Since the range is so wide for the y variable we see very little difference between these curves for the smaller values of x.

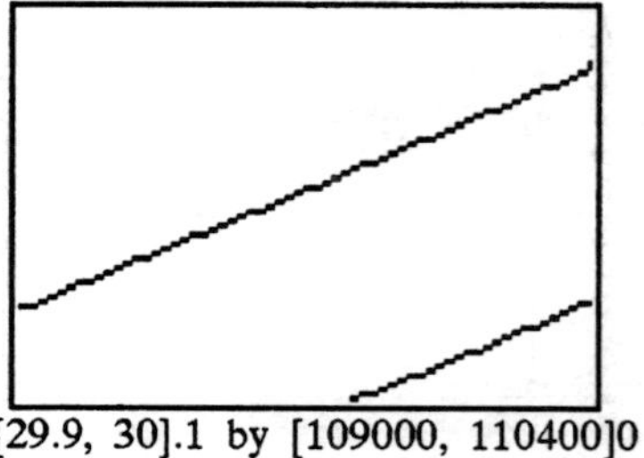
[29.9, 30].1 by [109000, 110400]0

Change the viewing window variables to [29.9, 30].1 by [109000, 110400]0 to see the difference between the graphs for the last tenth of the thirtieth year. We see there is little difference between result of monthly and daily compounded interest.

Try changing the viewing window to other settings to see these graphs. One suggestion is [25, 30]1 by [90000, 111000]0.

EXAMPLE 3 — *TEXTBOOK SECTION 3-2*

$4,570 is invested at 7.8% compounded monthly. (A) Make a chart showing the amount in the account at the end of each period for eight months. (B) How much interest is earned each month? (C) How much interest is earned overall? (D) How does the interest in the first month relate to that in each of the succeeding months? (E) Repeat Parts (A) through (D) under the conditions that the interest is withdrawn from the account at the end of each month. (F) Graph the future value as a function of time for each of the situations in this example. Why is the graph for the situation in Part (E) a horizontal line?

Solution (A) & (B): <u>TI-81, TI-82, and TI-85</u> Store the expression $P(1+i)^n$ in the calculator as a function in the calculator as P(1+I)^N. Store P×I as a separate function in the calculator. Store 4570 as P, .078÷12 as I, and 1 as N.

Calculate the second expression P×I. The result is $29.71.
Calculate the first expression $P(1+i)^n$. The result is $4599.71.
Store this result as P.

Calculate the second expression again. The result is $29.90
Calculate the first expression again. The result is $4629.60.
Store this result as P.

Repeat this procedure until the chart has been filled in.

Period Number	Principal at the start of the month	Amount at the end of the month	Amount of interest earned each period
1	4570	4599.71	29.71
2	4599.71	4629.60	29.90
3	4629.60	4659.70	30.09
4	4659.70	4689.98	30.29
5	4689.98	4720.47	30.48
6	4720.47	4751.15	30.68
7	4751.15	4782.03	30.88
8	4782.03	4813.12	31.08
			Sum = $243.12

TI-83 Use the built-in financial functions of the calculator to calculate the amount at the end of each month. Find the interest by subtraction.

Keystrokes	Explanation
[4570] [STO▶] [ALPHA] [A]	To store the principal at the beginning of the time period as A.
[2nd] [FINANCE] [1] <TVM Solver...>	To get the FINANCE CALC menu and TVM Solver.
[1] [ENTER]	To set N=1, the number of months.
[7.8] [÷] [12] [ENTER]	To set I%=8/12, the interest rate per month.
[(-)] [4570] [ENTER]	To set PV=-4570, as a cash outflow which is the present value deposited.
[0] [ENTER]	To set PMT=0 since there is no payment each period.
[ENTER]	It is not necessary to enter a value for FV since this is what we are looking for.
[1] [ENTER]	To set P/Y=1, the number of payment periods per month.
[1] [ENTER]	To set C/Y=1, the number of compounding periods per month.
[ENTER]	So the calculation of interest will be done at the end of the compounding period.

Keystrokes	Explanation
[▲] [▲] [▲] [ALPHA] [SOLVE]	Place the cursor on FV and solve for the future value. The future value for the first month is $4599.71.
[2nd] [QUIT]	To exit the finance menu.
[2nd] [FINANCE] [6] <tvm_FV> [-] [ALPHA] [A] [ENTER]	To find the amount of interest earned this period.
[2nd] [FINANCE] [6] <tvm_FV> [STO▸] [ALPHA] [A]	To store the principal for the next time period.

Repeat the above from the second line beginning [2nd] [FINANCE] ... and store 2 as N. Repeat for each value of N until the chart has been completed.

Solutions (C) and (D) can be found by changing the number of compounding periods per year, P/Y, to the appropriate value: P/Y=12 in Solution (C). P/Y=365 in Solution (D).

Solution (C): The sum of the interest amounts should be equal to the value of the expression $P(1+i)^n - P$ for P=4750, I=.078/12, and N=8. Make certain you still have the expression $P(1+i)^n$ stored in the calculator as P(1+I)^N. Store 4570 as P, .078÷12 as I, and 8 as N. Evaluate the expression. The result is $4813.12 Now subtract P. The result is $243.12. This is the same amount as the sum of the "Amount of interest earned each period" column in the table on the preceding page. [NOTE: The financial function of the TI-83 could be used as shown in Solution (B) and using N = 8.

Solution (D): The interest on each succeeding month is the interest from the previous month plus the interest on this amount. For example: The interest for month 2 is $29.71 + 29.71 \times \frac{.078}{12}$. This demonstrates that the interest also earns interest each succeeding month.

Solution (E): The chart is now:

Period Number	Principal at the start of the month	Amount at the end of the month	Amount of interest earned each period
1	4570	4599.71	29.71
2	4570	4599.71	29.71
3	4570	4599.71	29.71
4	4570	4599.71	29.71
5	4570	4599.71	29.71
6	4570	4599.71	29.71
7	4570	4599.71	29.71
8	4570	4599.71	29.71

Sum = $29.71x8 = $237.68

When the interest earned during a month is withdrawn at the end of the month, the principal remains the same each month and hence, the interest earned in each month is the same.

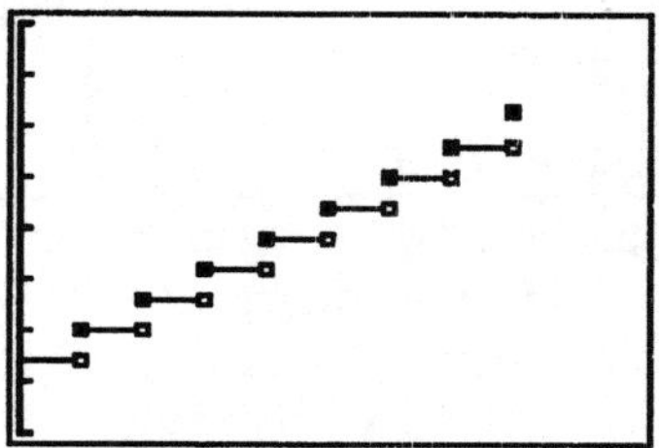

Solution (F): The expression for the first situation is $4570\left(1 + \frac{.078}{12}\right)^x$ for integer values of x so that the interest is calculated and added to the principal at the beginning of the next month. The graph is shown to the right. The graph can be displayed on the calculator by entering (4570 (1 + .078 ÷ 12) ^ iPart X) as Y1,
setting the graph mode to dot and using [0, 9.5]1 by [4500, 4900]0 on the TI-82 and TI-83, and [0, 12.6]1 by [4500, 4900]0 on the TI 85 and TI-86. iPart is found on the MATH menu. The open and closed dots are not displayed on the calculator. (See Appendix Section A-11, B-11, C-11, or C-11 to see how to get iPart on your calculator.)

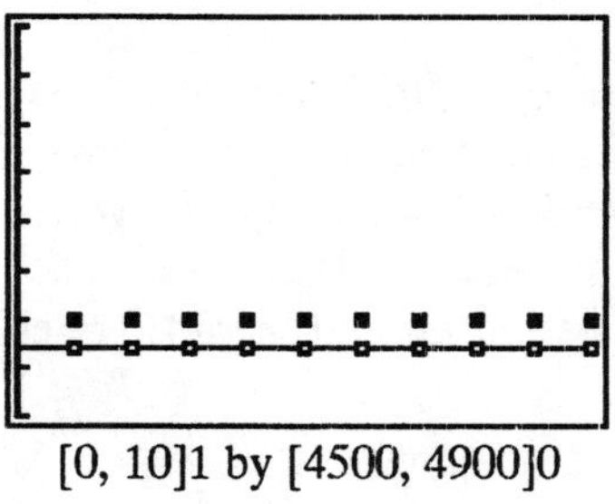

[0, 10]1 by [4500, 4900]0

In the second situation, the amount in the account is $4,570 for the entire month until the end of the month when the interest is calculated. The interest is added to the account at the beginning of the next month but then immediately withdrawn. This graph can be displayed using the same viewing window variables. Store 4570 (1 + .078 ÷ 12) (X = iPart X) as Y1 and 4570 ÷ (X ≠ iPart X) as Y2 where the = and ≠ signs are from the TEST menu and iPart is from the MATH menu. The open and closed dots are not displayed on the calculator.

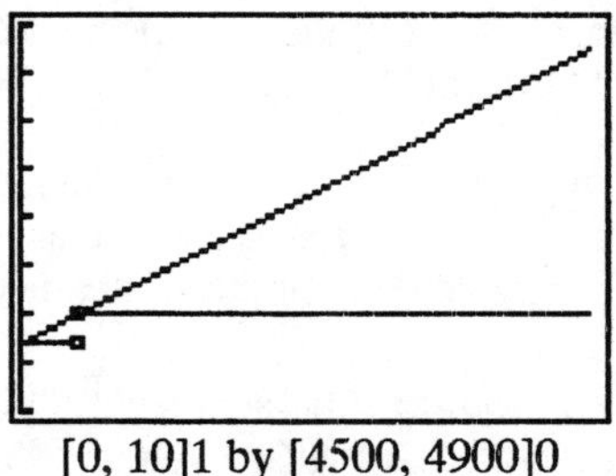

[0, 10]1 by [4500, 4900]0

We usually graph the above two functions as continuously increasing functions. To do this enter 4570 (1 + .078 ÷ 12) ^ X as Y1 in the calculator for the first situation. Enter 4570 ÷ (X < 1) as Y2 and (4570 (1 + .078 ÷ 12)) ÷ (X > 1) as Y3 and for the second situation. The inequality signs are found on the TEST menu. Set the viewing window dimensions at [0, 10]1 by [4500, 4900]0. Graph. The open and closed dots are not displayed on the calculators. (See graph to the right.)

Note that the future value starts at $4570 since both situations start with the same principal. The graph in the second situation is a horizontal line since the future value for each month is $4599.71 as the principal does not change each month since the interest is withdrawn from the account at the end of each month.

EXAMPLE 4 **TEXTBOOK SECTION 3-2**

Graph the growth of an investment of $5,000 at 10% compounded semiannually and the growth of an investment of $4,500 at 12% compounded daily. (A) Which investment would be better if you plan to leave the money invested for a period of 10 years? (B) How much time will have elapsed before the investments yield the same amount?

Solution (A): We must first store the appropriate expressions in the functions list in the calculator. Enter 5000 (1 + .10 ÷ 2) ^ (2 x X) as Y1 for the first investment. Enter 4500 (1 + .12 ÷ 365) ^ (365 x X) as Y2 for the second investment. Set the viewing window variables at [0, 10]1 by [5000, 25000]500. The graphs are shown to the right. The investment of $4500 at 12% is the better investment if money is left in the account for 10 years.

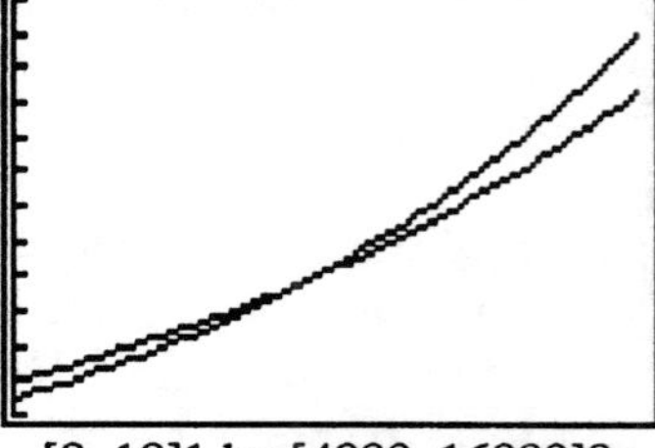
[0, 10]1 by [4000, 16000]0

Solution (B): Trace and zoom, or use other calculator methods, to find that the point of intersection occurs at x=4.7 years. Since the first investment calculates interest each 6-months, the investments would be said to be equal after 5 years.

EXAMPLE 5 **TEXTBOOK SECTION 3-3**

$150 is deposited at the end of each month in an account earning 5.6% interest compounded monthly. Deposits were started five years ago. How much interest was earned during the third year?

***Solution*:** We need to find the amount that was in the account at the end of the second year. Then we need to find the amount that was in the account at the end of the third year. The amount of interest earned during the third year is then given by

$$\left(\begin{array}{c}\text{the amount in the account}\\ \text{at the end of the 3rd year}\end{array}\right) - \left(\begin{array}{c}\text{the amount in the account}\\ \text{at the end of the 2nd year}\end{array}\right) - \left(\begin{array}{c}\text{the amount deposited}\\ \text{the third year}\end{array}\right)$$

To find the amount in the account at the end of the second year calculate

$$S = 150\,\frac{\left(1 + \frac{.056}{12}\right)^{24} - 1}{\frac{.056}{12}}.$$

Enter this into the calculator as 150 ((1 + .056 ÷ 12) ^ 24 - 1) ÷ (.056 ÷ 12). The result is $3,799.98.

To find the amount in the account at the end of the third year calculate

$$S = 150\,\frac{\left(1 + \frac{.056}{12}\right)^{36} - 1}{\frac{.056}{12}}.$$

The result is $5,865.25. The amount deposited during the third year is $150 x 12 = $1,800. Hence the total interest earned in the third year is $5,865.25 – 3,799.98 – 1,800 = $265.27.

<u>TI-83</u>

Keystrokes	Explanation
2nd | FINANCE | 1 <TVM Solver...>	To get the FINANCE CALC menu and TVM Solver.
24 | ENTER	To set N=24, the number of payment periods.
5.6 | ENTER	To set I%=5.6, the interest rate per year.
0 | ENTER	To set PV=0, as there is no present value at the start.
(-) | 150 | ENTER	To set PMT=-150 since the payment is $150/period.
ENTER	It is not necessary to enter a value for FV since this is what we are looking for.
12 | ENTER	To set P/Y=12, the number of payment periods per year.
12 | ENTER	To set C/Y=12, the number of compounding periods per year.
ENTER	So the calculation of interest will be done at the end of the compounding period.
▲ | ▲ | ▲ | ALPHA | SOLVE	Place the cursor on FV and solve for the future value. The future value is $3799.98.
2nd | QUIT	To exit the finance menu.
2nd | FINANCE | 6 <tvm_FV> STO▸ | ALPHA | A	To store the amount in the account at the end of 2 years.

Repeat using 36 as *N* to find the amount in the account at the end of 3 years.

Keystrokes	Explanation
2nd | QUIT	To exit the finance menu.
2nd | FINANCE | 6 <tvm_FV> - | ALPHA | A | - | 150 | ENTER	To find the amount of interest earned during the third year.

EXAMPLE 6 — ***TEXTBOOK SECTION 3-3***

A sinking fund has been accumulating money for five years. $250 is deposited at the end of each month. Currently there is $17353.15 in the account. What is the interest rate?

Solution: The formula is $S = R\,\dfrac{(1+i)^n - 1}{i}$. We are looking for $r = 12i$. We wish to find i from this equation. However, the formula cannot be solved for i algebraically.

i can be approximated graphically. Substituting the numbers into the equation for S, R, and n, we have $17353.15 = 250\,\dfrac{(1+i)^{60} - 1}{i}$. Store the left side of this equation as a function as 17353.15 in the calculator. Store the right side of this equation as a second function as: 250 ((1 + X) ^ 60 - 1) ÷ X where X represents i.

We now need to determine values for the viewing window variables. We know i will be between 0 and 1 since it is the monthly interest rate. So the viewing window variables for the horizontal axis will be [0, 1]1.

Since the first function stored is 17353.15 we must scale the y axis to accommodate this large number. Hence graph using [0, 1].1 by [0, 20000]1000. However, when this viewing window is used, nothing of the second function is seen on the graph. Hence we must adjust the variables. We suspect that perhaps the first screen dimensions are too big. Changing the viewing window to [0, .1].01 by [0, 20000]1000 gives the graph on the left below (see next page).

Using trace we see that the point of intersection is near (.005, 17500). Zoom or change the range to [0, .05].01 by [17000, 18000]1000, to get the graph on the right. (Note: It may take the calculator a little time to graph.)

Continue tracing and zooming, or changing the range, until you find $x \approx 0.00483$. This is the value of i. The rate we are seeking is $12i$ = .058 or 5.8%.

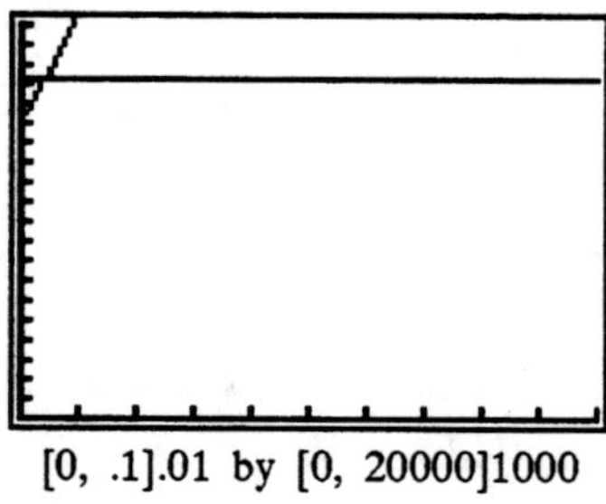

[0, .1].01 by [0, 20000]1000

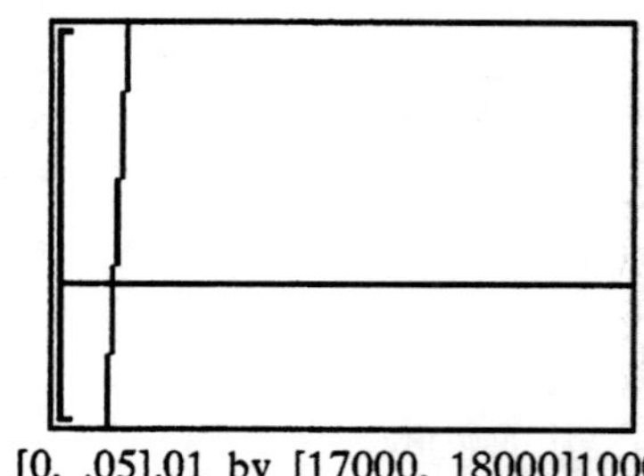

[0, .05].01 by [17000, 18000]1000

TI-83

Keystrokes	Explanation
[2nd] [FINANCE] [1] <TVM Solver...>	To get the FINANCE CALC menu and TVM Solver.
[60] [ENTER]	To set N=60, the number of payment periods.
[ENTER]	You do not need to enter any particular value for I% since this is what we are looking for.
[0] [ENTER]	To set PV=0, as there is no present value at the start.
[(-)] [250] [ENTER]	To set PMT=-250 since the payment is $250/period.
[17353.15] [ENTER]	To set FV=17353.15, the future value.
[12] [ENTER]	To set P/Y=12, the number of payment periods per year.
[12] [ENTER]	To set C/Y=12, the number of compounding periods per year.
[ENTER]	So the calculation of interest will be done at the end of the compounding period.
[▲] [▲] [▲] [▲] [▲] [▲]	Place the cursor on I% and solve for the interest rate.
[ALPHA] [SOLVE]	
	The interest rate is 5.8%.

EXAMPLE 7 ***TEXTBOOK SECTION 3-4***

Suppose you have purchased a car for $11,500. The rate of financing is 9.8% compounded monthly.

(A) What are your monthly payments if you finance for one year?

(B) Make an amortization schedule.

(C) How much interest do you pay if you do not pay off the loan early?

(D) Revise the table to reflect the situation if you made $1,500 payments instead those calculated in Part (A).

(E) How much interest is paid in the situation of Part (D)?

(F) Find the payment number after which no more than one-half of the original unpaid balance is due.

Solution (A): Since we are dealing with money in this problem, it may be helpful to set the number of decimal places in the calculator to two. (See Appendix Section A-17, B-17, C-17 and D-17 of this manual.)

The formula for finding the payment is $P = 11500 \dfrac{\frac{.098}{12}}{1 - \left(1 + \frac{.098}{12}\right)^{-12}}$. Enter this into the calculator as: 11500 ((.098 ÷ 12) ÷ (1 - (1 + .098 ÷ 12) ^- 12)).

The result of the calculation is 1009.96. The payments are $1,009.96.

TI-83

Keystrokes	Explanation
[2nd] [FINANCE] [1] <TVM Solver...>	To get the FINANCE CALC menu and TVM Solver.
[12] [ENTER]	To set N=12, the number of payment periods.
[9.8] [ENTER]	To set the interest at 9.8%.
[11500] [ENTER]	To set PV=11500, the present value (cost of the car).
[ENTER]	Do not need to set this to a particular value since this is what we are looking for.
[0] [ENTER]	To set FV=0, the future value.
[12] [ENTER]	To set P/Y=12, the number of payment periods per year.
[12] [ENTER]	To set C/Y=12, the number of compounding periods per year.
[ENTER]	So the calculation of interest will be done at the end of the compounding period.
[▲] [▲] [▲] [▲]	Place the cursor on PMT and solve for the payment.
[ALPHA] [SOLVE]	The result is negative since this is a payment. The payment is $1009.96.

Solution (B): The amortization schedule can be found as follows:

Store the functions $P \times I$, $A - P \times I$, and $P - (A - P \times I)$ as three separate functions in the calculator. Store 11500 as P, .098÷12 as I, and 1009.96 as A.

Evaluate the three stored functions. The results are the interest, unpaid balance reduction, and unpaid balance, respectively. The results of the first calculation are 93.92, 916.04, and 10583.96. Store Ans as P. Repeat the instructions in this paragraph, recording the results as you go until the following table is complete.

Payment Number	Payment	Interest on Unpaid Balance	Unpaid Balance Reduction	New Unpaid Balance
1	1009.96	93.92	916.04	10583.96
2	1009.96	86.44	923.52	9660.43
3	1009.96	78.89	931.07	8729.37
4	1009.96	71.29	938.67	7790.70
5	1009.96	63.62	946.34	6844.36
6	1009.96	55.90	954.06	5890.30
7	1009.96	48.10	961.86	4928.44
8	1009.96	40.25	969.71	3958.73
9	1009.96	32.33	977.63	2981.10
10	1009.96	24.35	985.61	1995.48
11	1009.96	16.30	993.66	1001.82
12	1009.96	8.18	1001.78	.04

Note that the unpaid balance is not 0. This is because of rounding that takes place during the calculations. This amount probably would be added to your last payment.

TI-83

The unpaid balance column can be filled in by using the finance feature of the calculator as shown below:

Keystrokes	Explanation
[2nd] [FINANCE] [1] <TVM Solver...>	To get the FINANCE CALC menu and TVM Solver.
[1] [ENTER]	To set N=1, the number of payment periods.
[9.8] [ENTER]	To set I%=9.8, the interest rate per year.
[11500] [ENTER]	To set PV=15000, as this is the amount owed at the beginning of the first period.
[(-)] [1009.96] [ENTER]	To set PMT=-1009.96, the payment each period.
[ENTER]	This does not need to be set as this is what we are looking for, the future value at the end of the first payment period.
[12] [ENTER]	To set P/Y=12, the number of payment periods per year.
[12] [ENTER]	To set C/Y=12, the number of compounding periods per year.
[ENTER]	So the calculation of interest will be done at the end of the compounding period.
[▲] [▲] [▲] [ALPHA] [SOLVE]	Place the cursor on FV and solve for the future value. The result is negative since this is what you still owe. The result is -10583.96. The new unpaid balance is $10,583.96.

[▲] [▲] [2nd] [FINANCE] [(-)] — To store the unpaid balance as the new present value.

[6] <TVM_FV>

[ENTER] — Leave the payment at $1009.96

[ENTER] [ALPHA] [SOLVER] — To get the next unpaid balance.

The interest on the unpaid balance for the next period is found by multiplying the new unpaid balance by the interest rate for one period, .098÷12 = .00081667.

The unpaid balance reduction is found by subtracting the interest from the payment.

Solution (C): The amount of interest is $1009.96×12 - $11500 = $619.52.

Solution (D): The payment schedule is:

Payment Number	Payment	Interest on Unpaid Balance	Unpaid Balance Reduction	New Unpaid Balance
1	1500.00	11500.00x.098/12= 93.92	1500-93.92= 1406.08	11500.00-1406.08 = 10093.92
2	1500.00	82.43	1417.57	8676.35
3	1500.00	70.86	1429.14	7247.21
4	1500.00	59.19	1440.81	5806.40
5	1500.00	47.42	1452.58	4353.82
6	1500.00	35.36	1464.44	2889.38
7	1500.00	23.60	1476.40	1412.98
8	1500.00	11.54	1488.46	-75.48
9				
10				
11				
12				

We notice that the last value in the "New Unpaid Balance" column is -75.48. This means that the last payment need only be 1500.00-75.48 or $1,424.52.

Solution (E): The amount of interest is (1500.00×7+1424.52) - 11500 = 424.52. This illustrates the value in making payments in excess of those required if your contract will allow you to do so.

Solution (F): Half of the original unpaid balance of \$11,500 is \$5,750. We wish to find x so that $5750 = 1009.96 \dfrac{1 - \left(1 + \dfrac{.098}{12}\right)^{-(12-x)}}{\left(\dfrac{.098}{12}\right)}$.

We see from the table in Part (A) that this occurs between the sixth and seventh payment. Enter 5750 as Y1 in the calculator. Enter 1009.96 ((1 - (1 + .098 ÷ 12) ^ - (12 - X)) ÷ (.098 ÷ 12)) as Y2 in the calculator. Set the viewing window variables to [0, 8]1 by [4000, 7000]1000. Graph. The result is shown to the right.

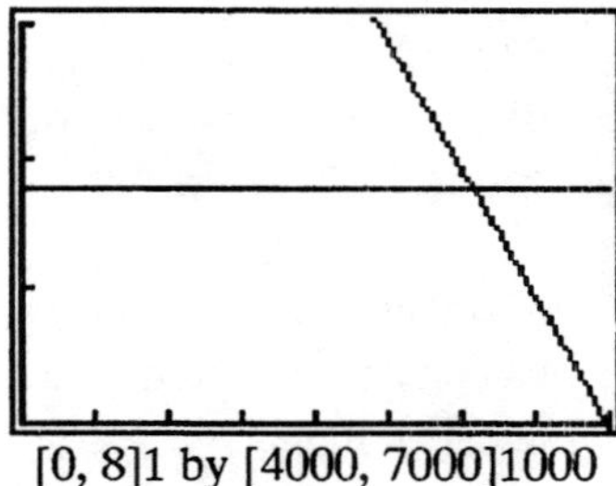
[0, 8]1 by [4000, 7000]1000

Trace and zoom, or use other calculator methods, to find the x coordinate of the point of intersection. Since this is not an integer, we round up. Hence the number of payments needed is seven. The unpaid balance after the seventh payment will be a little less than one half of the original unpaid balance.

Exercise Set 3

1. \$43,865 is invested in an account earning simple interest of 4.96%. How much interest will be earned if the money is left in the account for one year? Two years? Five years?

2. \$8,392 is invested in an account earning simple interest of 7.83%. How much interest will be earned if the money is left in the account for one year? Two years? Five years?

3. \$6,925 is invested at 6.5% compounded quarterly. (A) Make a chart showing the amount in the account at the end of each quarter for 18 months. Include a column on the chart showing how much interest is earned each quarter. (B) Make a chart showing the amount of the investment at the end of each quarter for 18 months if the interest is withdrawn after it is calculated. (C) Make a graph comparing these two situations.

4. \$200 is invested at 5.8% compounded monthly. (A) Make a chart showing the amount in the account at the end of each month for 9 months. Include a column on the chart showing how much interest is earned each quarter. (B) Make a chart showing the amount of the investment at the end of each quarter for 18 months

if the interest is withdrawn after it is calculated. (C) Make a graph comparing these two situations.

5. $50 is deposited each month in an account earning 4.8% interest compounded monthly. Deposits were started five years ago. How much interest was earned during the third year?

6. $375 is deposited each month in an account earning 7.2% interest compounded daily. Deposits were started five years ago. How much interest was earned during the third year?

7. A sinking fund has been accumulating money for five years. $200 is deposited each month. Currently there is $13532 in the account. What is the interest rate?

8. You bought a new car for $12,500. Your payments are $743.25 for 18 months. Find the interest rate?

9. A stereo system has been purchased for $3500. The rate of financing is 9.8% compounded monthly. What are your payments if you finance for one year? Make an amortization schedule. How much interest do you pay?

10. Repeat Example 5 above if you finance at 8.5% for 18 months. Do you pay more interest under this plan?

NOTES

Chapter 4

Systems of Linear Equations; Matrices

This chapter contains examples using the calculator to illustrate:

- Solving of systems of equations in two variables using graphing
- Solving of systems of equations using Gauss-Jordan reduction
- Solving systems of equations using the inverse coefficient matrix
- Solving problems using matrix operations

EXAMPLE 1 ***TEXTBOOK SECTION 4-1***

Given the system of equations $\begin{aligned} 11x + 2y &= 1 \\ 2x + 5y &= -24 \end{aligned}$.

(A) Solve to two decimal places using graphing techniques.

(B) Conjecture, then check, the effect on the solution of changing the 11 in the first equation to 20.

(C) Conjecture, then check, the effect on the solution of changing the 1 in the first equation of the original system to 6.

Solution (A): Each of these equations must be solved algebraically for y in order to enter it into the calculator.

The system becomes:	Store in the calculator as two separate functions:
$y = \frac{1-11x}{2}$	(1 - 11 X) ÷ 2
$y = \frac{-24-2x}{5}$	(-24 - 2 X) ÷ 5

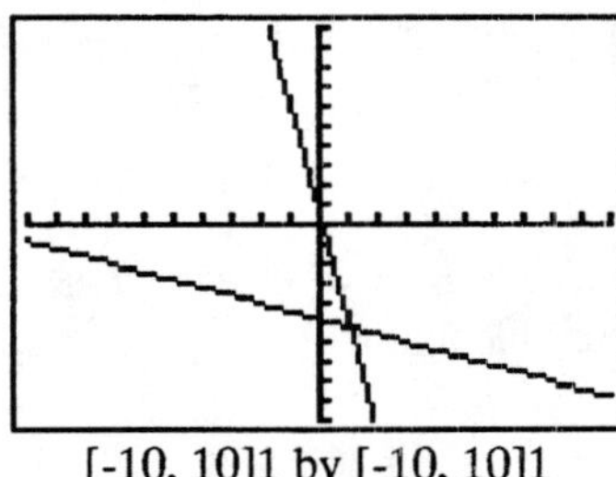

[-10, 10]1 by [-10, 10]1

Graph the lines using [-10,10]1 by [-10,10]1. Trace and zoom, or use other calculator techniques, to find the coordinates of the point of intersection. The solution is $x = 1.04$, $y = -5.22$.

Solution (B): Change the 11 in the first equation to 20. Store this in the calculator as Y3 as (1 - 20 X) ÷ 2. Changing the 11 in the first equation to 20 changes the slope of the line from $-\frac{11}{2} = -5.5$ to $-\frac{20}{2} = -10$. This means the line has steeper slope but the same y intercept.

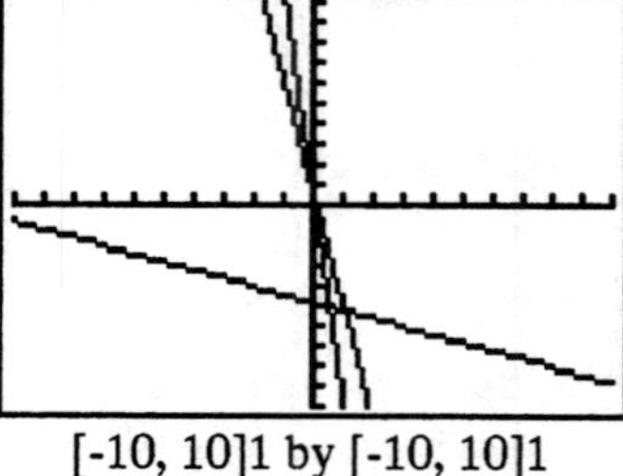

[-10, 10]1 by [-10, 10]1

Hence the x coordinate of the point of intersection will move to the left and the y coordinate will move upward since the slope of the second line is negative. The coordinates of the point of intersection are (.55, -5.02).

(All three lines are graphed to the right.)

Solution (C): Change the 1 to 6 in the original first equation. Store this in the calculator as Y3 as (6 - 11 X) ÷ 2. Changing the 1 to 6 in the original first equation changes the y intercept from $\frac{1}{2}$ to $\frac{6}{2} = 3$. This shifts the first line upward (or to the right). Hence the point of intersection will move to the right and downward. The x coordinate will be greater and the y coordinate will be less than the coordinates of the original intersection point. The solution is $x = 1.53$, $y = -5.41$.

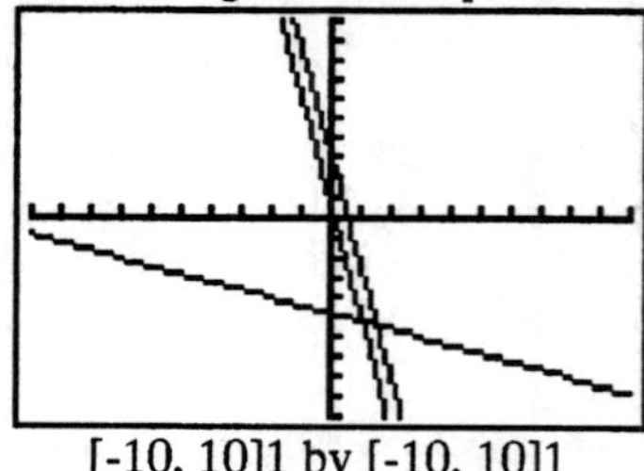

[-10, 10]1 by [-10, 10]1

EXAMPLE 2 ***TEXTBOOK SECTION 4-1***

Solve
$$\begin{aligned} 9x - 3y &= 24 \\ 11x + 2y &= 1 \\ 2x + 5y &= -24 \end{aligned}$$
using graphing techniques.

Solution: Each of these equations must be solved algebraically for y in order to enter it into the calculator.

The system becomes:

$$y = \frac{24-9x}{-3}$$

$$y = \frac{1-11x}{2}$$

$$y = \frac{-24-2x}{5}$$

Store in the calculator as three separate functions:

(24 - 9 X) ÷ -3

(1 - 11 X) ÷ 2

(-24 - 2 X) ÷ 5

Graph these three lines using [-10, 10]1 by [-10, 10]1. The lines appear to intersect in a single point. Zoom or change the viewing window variables to [.55, 1.55]1 by [-5.8, -4.6]1 to see that the three lines do not intersect in a single point (see graphs below). Consequently there is no one ordered pair that satisfies all three equations at the same time. Hence there is no solution to this system of equations.

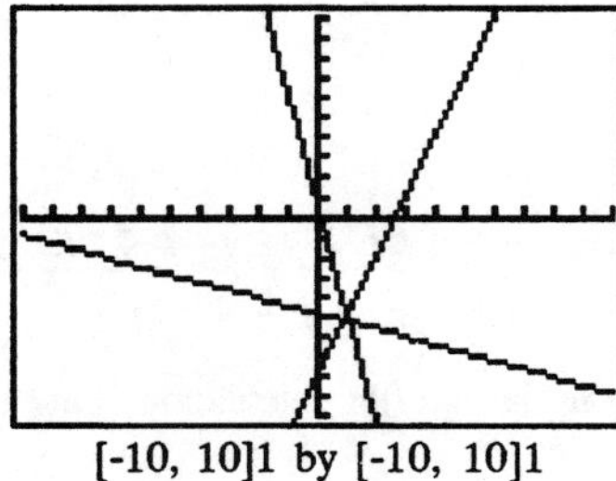
[-10, 10]1 by [-10, 10]1

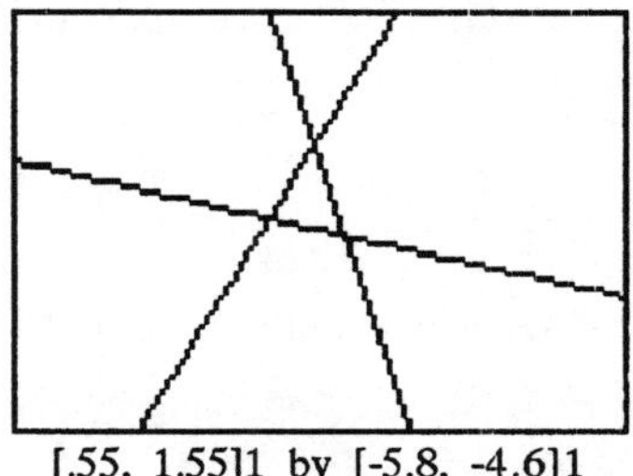
[.55, 1.55]1 by [-5.8, -4.6]1

EXAMPLE 3 ***TEXTBOOK SECTION 4-2 & 4-3***

Given the system of equations:

$$\begin{aligned} 7x - 2y + z &= 3 \\ 3x + y - z &= 6 \\ x - 3y + 2z &= 0 \end{aligned}$$

(A) Solve to two decimal places using Gauss-Jordan Elimination.

(B) Conjecture, then check, how the solution for x would be changed if the 3 in the first equation was changed to a 0.

(C) Conjecture, then check, what the solution to the system would be if both the 3 in the first equation and the 6 in the second equation are changed to 0.

Solution (A): Enter the augmented coefficient matrix into the calculator. The particular keystrokes for your calculator are described in Appendix Section A-12, B-12, C-12 or D-12 Example 4 of this manual. Use the built-in features of your calculator to perform row operations to get the reduced form.

$$\left[\begin{array}{ccc|c} 7 & -2 & 1 & 3 \\ 3 & 1 & -1 & 6 \\ 1 & -3 & 2 & 0 \end{array}\right] \sim \left[\begin{array}{ccc|c} 1 & 0 & 0 & -1 \\ 0 & 1 & 0 & -19 \\ 0 & 0 & 1 & -28 \end{array}\right].$$

Hence the solution to this system is $x=-1$, $y=-19$, and $z=-28$.

Solution (B): If the 3 is changed to a 0 the value for x will be less than -1. In fact, the solution for x is - 2.

Solution (C): The solution would be $x=0$, $y=0$, and $z=0$ since there would be no values added or subtracted in the constant column.

EXAMPLE 4 ***TEXTBOOK SECTION 4-4 & 4-5***

Given the matrices below find: (A) AB (B) BA (C) $A + A^{-1}$ (D) 2.18A (E) A^2 and A^3. Round all resulting numbers to two decimal places after calculations have been completed.

$$A = \begin{bmatrix} .28 & .31 & .45 \\ .06 & -.02 & -.40 \\ .09 & -.42 & -.93 \end{bmatrix} \qquad B = \begin{bmatrix} .47 \\ -6.38 \\ .81 \end{bmatrix}$$

Use two decimal place accuracy in all answers.

Solution: Store matrices A and B in the calculator. (See Appendix Section A-12, B-12, C-12 or D-12 of this manual.)

(A) Find the product AB. The result is $\begin{bmatrix} -1.48 \\ -.17 \\ 1.97 \end{bmatrix}$.

(B) Find the product BA. An error message appears on the calculator screen. This is because the dimensions are not correct for multiplication. You cannot multiply a 3x1 matrix by a 3x3 matrix.

(C) Find the sum $A + A^{-1}$. The result is $\begin{bmatrix} 3.51 & -1.84 & 2.94 \\ -.37 & 6.49 & -3.41 \\ .60 & -3.57 & -.41 \end{bmatrix}$.

(D) Find 2.18A . The result is $\begin{bmatrix} .61 & .68 & .98 \\ .13 & -.04 & -.87 \\ .20 & -.92 & -2.03 \end{bmatrix}$.

(E) Find A^2 by recalling matrix A and multiplying it by itself or using the $\boxed{x^2}$ key.

The result is $\begin{bmatrix} .14 & -.11 & -.42 \\ -.02 & .19 & .41 \\ -.08 & .43 & 1.07 \end{bmatrix}$.

Now recall the answer using $\boxed{\text{2nd}}$ $\boxed{\text{ANS}}$ and multiply it by A.

The result is $\begin{bmatrix} -.01 & .22 & .49 \\ .04 & -.18 & -.46 \\ .10 & -.49 & -1.21 \end{bmatrix}$.

EXAMPLE 5 — *TEXTBOOK SECTION 4-4*

The time requirements for the manufacture of slalom and trick water skis is given by :

	Fabricating department	Finishing department	
Trick ski	6.0 hr	1.5 hr	= A
Slalom ski	4.0 hr	1.0 hr	

Now suppose the company has two manufacturing plants, X and Y, in different parts of the country and that their hourly rates for each department are given in the following matrix:

	Plant X	Plant Y	
Fabricating department	\$8/hr	\$7/hr	= B
Finishing department	\$6/hr	\$4/hr	

(A) Find the total labor costs for each ski at each factory.

(B) Conjecture, then check, how the total labor costs change if the hourly rates at Plant X in the fabricating department and the finishing department increased to \$8.50/hr and \$6.35/hr respectively?

Solution (A): Enter the two matrices into the calculator. (See Appendix Section A-12, B-12, C-12 or D-12 of this manual.) Find AB. The result is $\begin{bmatrix} \$57 & \$48 \\ \$38 & \$32 \end{bmatrix}$.

Solution (B): A good conjecture would be that the total labor costs would increase since both the hourly rates increased. Change the matrix elements in matrix B as indicated above and recalculate. The result is now is $\begin{bmatrix} \$60.53 & \$48 \\ \$40.35 & \$32 \end{bmatrix}$. The labor costs at Plant X increase $3.53 per ski for the trick ski and $2.35 per ski for the slalom ski.

EXAMPLE 6 ***TEXTBOOK SECTION 4-6***

Solve to two decimal places using inverse matrix methods:

$$\begin{aligned} 7x - 2y + z &= 3 \\ 3x + y - z &= 6 \\ x - 3y + 2z &= 0 \end{aligned}$$

Solution: Enter the coefficient matrix into the calculator. The particular keystrokes for your calculator are described in Appendix Section A-12, B-12, C-12 or D-12 Example 2 of this manual. Use the built-in features of your calculator to get the inverse matrix shown below.

$$A^{-1} = \begin{bmatrix} .3333 & -.3333 & -.3333 \\ 2.3333 & -4.3333 & -3.3333 \\ 3.3333 & -6.3333 & -4.3333 \end{bmatrix}$$

Find the matrix product of A^{-1} and B:

$$\begin{bmatrix} .3333 & -.3333 & -.3333 \\ 2.3333 & -4.3333 & -3.3333 \\ 3.3333 & -6.3333 & -4.3333 \end{bmatrix} \begin{bmatrix} 3 \\ 6 \\ 0 \end{bmatrix} = \begin{bmatrix} -1 \\ -19 \\ -28 \end{bmatrix}$$

. Hence $x = -1$, $y = -19$, and $z = -28$.

Exercise Set 4

1. Solve $\begin{aligned} 7x - 5y &= 14 \\ 11x + 2y &= 1 \\ 2x + 5y &= -10 \end{aligned}$ using graphing techniques.

2. Solve $\begin{aligned} 7x - 5y &= 14 \\ 11x + 2y &= 1 \\ 2x + 3.5y &= -7.5 \end{aligned}$ using graphing techniques.

3. Conjecture, then check, how the solution to Problem 1 above will change if the constant in the first equation is 16 instead of 14.

4. Conjecture, then check, how the solution to Problem 2 above will change if the coefficient on x in the first equation is 6 instead of 7.

5. Solve to two decimal places using Gauss-Jordan reduction:

$$\begin{aligned} 6x - 3y + 2z &= 3 \\ 3x + 2y - 3z &= 6.1 \\ x - 3y + z &= 5.2 \end{aligned}$$

6. Solve to two decimal places using Gauss-Jordan reduction:

$$\begin{aligned} 5x + 3y + 2z &= 4.5 \\ 3.1x - 2y + z &= 6.2 \\ 3x - 3.7y + 2.2z &= 9 \end{aligned}$$

7. Solve to two decimal places using inverse matrix methods:

$$\begin{aligned} 5x + 3y + 2z &= 4.5 \\ 3.1x - 2y + z &= 6.2 \\ 3x - 3.7y + 2.2z &= 9 \end{aligned}$$

8. Solve to two decimal places using inverse matrix methods:

$$\begin{aligned} 6x - 3y + 2z &= 3 \\ 3x + 2y - 3z &= 6.1 \\ x - 3y + z &= 5.2 \end{aligned}$$

9. Repeat all parts of Example 5 in this chapter of the manual if the hourly rate for the fabricating department at Plant X begins at $7.50/hr instead of $8/hr. Again, describe what happens when Plant X costs rise to $8.50/hr and $6.35/hr for fabricating and finishing, respectively.

10. Suppose a third type of ski was to be manufactured needing 8 hours in the fabricating department and 2 hours in the finishing department. Repeat Example 5 in this manual above under these conditions.

NOTES

Chapter 5

Linear Inequalities and Linear Programming

This chapter contains examples using the calculator to illustrate:

- Solving of systems of inequalities in two variables using graphing
- Finding corner points graphically
- Solving a linear programming problem in two variables using graphing
- Solving linear programming problems using matrix methods

EXAMPLE 1 ***TEXTBOOK SECTION 5-1***

Consider the system of linear inequalities listed below:

$$\begin{aligned} 2.3x + y &\geq 10 \\ 3x + y &\leq 14 \\ 1.8x + 4.8y &\leq 48.3 \\ x &\geq 0 \\ y &\geq 0 \end{aligned}$$

(A) Solve the system graphically.

(B) Find the corner points.

(C) Conjecture, then check, the effect on the solution if the 10 in the first inequality was changed to 5.

(D) Conjecture, then check, the effect on the solution if the 2.3 in the original system of inequalities was changed to a 1.

Solution (A): Step 1 Change the inequality signs to equal signs and graph the lines. This results in graphing the lines:

$$y = 10 - 2.3x$$
$$y = 14 - 3x$$
$$y = \frac{48.3 - 1.8x}{4.8}$$
$$x = 0$$
$$y = 0$$

Store the first three of these expressions in the calculator as separate functions: 10 - 2.3 X, 14 - 3 X, and (48.3 - 1.8 X) ÷ 4.8 . Graph using [0,15]1 by [0,15]1. We know the graph is in the first quadrant because of the last two inequalities.

Step 2 Change the expressions stored as Y1, Y2, and Y3 into statements with the appropriate inequality sign. (See Appendix Section A-5, B-5, C-5 or D-5 of this manual.) To do this use the INS function on the calculator and get the inequality signs from the TEST menu. The expressions are now stored in Y1, Y2, and Y3 as: Y ≥ 10 - 2.3 X, Y ≤ 14 - 3 X, and Y ≤ (48.3 - 1.8 X) ÷ 4.8.

Step 3 Choose a point not on any of the lines. We will use (1, 1). Enter 1 for X and 1 for Y in the calculator. Evaluate the inequalities by recalling each of the expressions stored in the calculator. If the result is 1 the statement is true and the point is in the solution region of that inequality. If the result in the calculator is 0, then the statement is not true and the point is not in the solution region. In this case we want the side of the line on the opposite side from (1, 1).

We see that the first inequality is false for $x=1$ and $y=1$. The second and third inequalities are true. So we want the region above the first line and below the second and third lines. This is shown in the figure to the right.

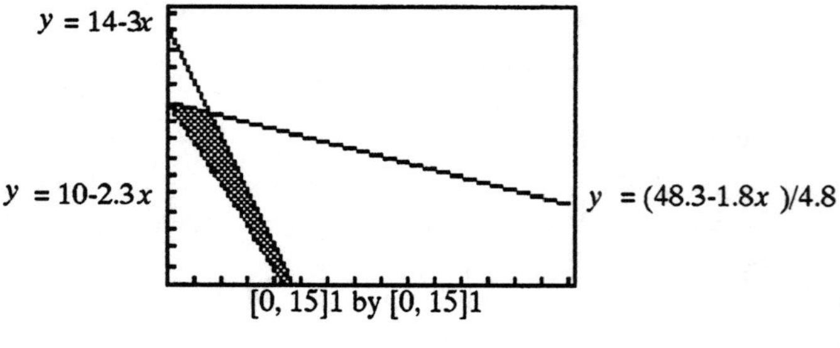

The solution of this system of linear inequalities is the set of ordered pairs corresponding to points in the shaded region and including the line segments that bound it. The shading cannot be done on all graphing calculators. It is shown here to indicate the solution region.

Solution (B): Change the expressions stored as Y1, Y2, and Y3 back to those in Step 1 above by deleting the Y and the inequality signs. Graph the lines. The corner points are found by tracing and zooming, or using other calculator methods (see Appendix Section A-7, B-7, C-7 or D-7 of this manual). The two corner points on the y axis are (0, 10) and (0, 10.06). The two corner points on the x axis are (4.35, 0) and (4.67, 0). The other corner point is (1.50, 9.50).

Solution (C): If the first inequality was $2.3x + y \geq 5$, the y intercept of the boundary line would be 5 instead of 10. This would shift the line downward. Since the slope of the line would be the same, the x intercept would shift to the left.

This situation is shown in the graph to the right. The corner points are now (0.00, 5.00), (0.00, 10.06), (2.17, 0.00), (4.67, 0.00) and (1.50, 9.50).

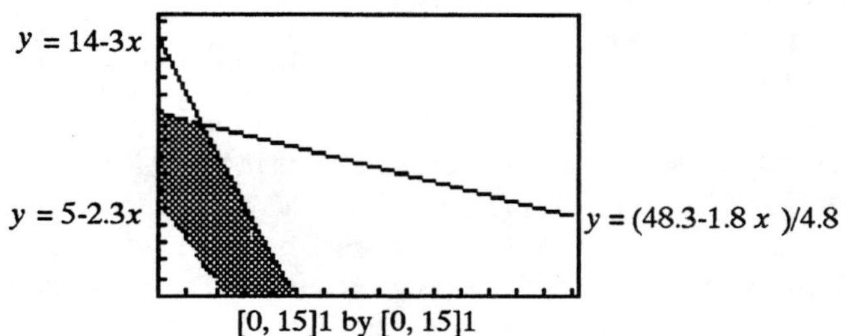

Solution (D): If the first inequality was $x + y \geq 10$, the slope of this boundary line is not as steep as that of the original boundary line. The y intercept remains the same. Hence the x intercept moves to the right. In fact, the x intercept is now 10. Since the x intercept of $y=14-3x$ is 4.67, the corner point is now the intersection of the two lines $y=14-3x$ and $y=10-x$.

The corner points of the solution region are now (0.00, 5.00), (0.00, 10.06), (2.00, 8.00), and (1.50, 9.50).

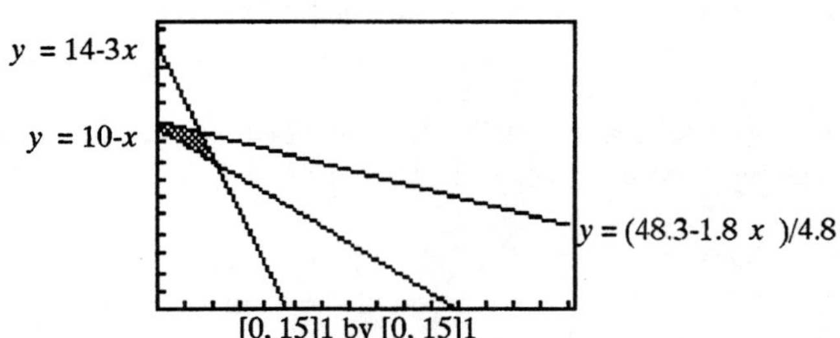

EXAMPLE 2 ***TEXTBOOK SECTION 5-2***

Maximize and minimize the function $z = 1.2x + y$ over the region graphed in Example 1 above.

Solution: We wish to see what the objective function line looks like. Graph the line $1.2x + y = 15$. (The value of 15 was chosen because the largest y intercept was 10.06.) Enter this in the calculator as 15-1.2X. Also graph 14-1.2X, 13-1.2X, 7-1.2X and 3-1.2X on the graph.

You may not be able to get all of the lines graphed on the screen at the same time. (See Appendix Sections A-6, B-6, C-6 or D-6 of this manual.) Sketch a graph on paper and add the parallel lines.

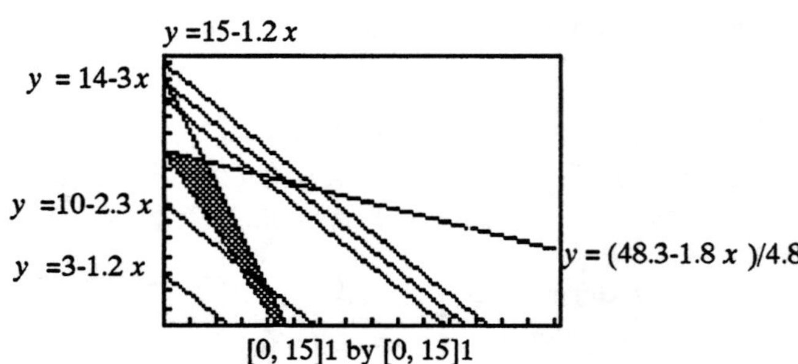

We see that as we decrease the constant value in the objective function line, the first corner point encountered is (1.50, 9.50). This will be the point that gives the maximum value to z. Similarly, we see that the smallest value for z will occur at the corner point (4.35, 0). Evaluating z at these points results in: A maximum value of z occurs at (1.5, 9.5). The maximum value of z is 11.3. A minimum value of z occurs at (4.35, 0). The minimum value of z is 5.22.

EXAMPLE 3 — *TEXTBOOK SECTION 5-4*

A manufacturer of lightweight mountain tents makes a standard model and an expedition model for national distribution. Each standard tent requires 1 labor hour from the cutting department and 3 labor hours from the assembly department. Each expedition tent requires 2 labor hours from the cutting department and 4 labor hours from the assembly department. The maximum labor hours available per day in the cutting department and the assembly department are 32 and 84, respectively.

If the company makes a profit of \$50 on each standard tent and \$80 on each expedition tent, how many tents of each type should be manufactured each day to maximize the total daily profit (assuming all tents can be sold)?

Solution: The mathematical model for this problem is:

$$\begin{aligned} \text{Maximize} \quad P &= 50x + 80y \\ \text{Subject to} \quad x + 2y &\le 32 \\ 3x + 4y &\le 84 \\ x,\ y &\ge 0 \end{aligned}$$

This problem could be done graphically since there are only two variables. However, we will use the Simplex Method.

The initial system is:

$$
\begin{aligned}
x + 2y + s_1 \quad &= 32 \\
3x + 4y \quad + s_2 \quad &= 84 \\
-50x - 80y \quad + P &= 0
\end{aligned}
$$

So the initial tableau is:

	x_1	x_2	S_1	S_2	P	
S1	1	2	1	0	0	32
S2	3	4	0	1	0	84
P	-50	-80	0	0	1	0

Use matrix row operations to perform the pivot operations. (See Appendix Section A-12, B-12, C-12 or D-12 of this manual.) The final tableau is:

	x_1	x_2	S_1	S_2	P	
S1	0	1	1.5	-.5	0	6
S2	1	0	-2	1	0	20
P	0	0	20	10	1	1480

So the maximum profit of $1480 is achieved if 20 standard tents and 6 expedition tents are manufactured and sold. Both slack variables have a value of zero. To see this, substitute the values for x and y into the slack variable equations. This means that all the hours available in the two departments are used.

Exercise Set 5

1. Solve the following system of linear inequalities graphically, and find the corner points:

$$
\begin{aligned}
4.5x + 2.1y &\geq 15 \\
3x + 7.2y &\geq 18.4 \\
1.8x + 4.8y &\leq 48.3 \\
x &\geq 0 \\
y &\geq 0
\end{aligned}
$$

2. Solve the following system of linear inequalities graphically, and find the corner points:

$$\begin{aligned} 5.5x + 3.1y &\geq 13.5 \\ 2x + 6.2y &\leq 19.2 \\ 1.8x + 4.8y &\leq 48.3 \\ x &\geq 0 \\ y &\geq 0 \end{aligned}$$

3. Maximize and minimize the function $z = 8.2x + 2.1y$ over the region graphed in Problem 1 above.

4. Maximize and minimize the function $z = 3.1x - y$ over the region graphed in Problem 2 above.

5. Repeat the first part of Example 3 above if the profit on standard tents is $90 and on expedition tents $70.

6. Repeat Example 3 if a new type of tent is introduced requiring 3 labor hours from the cutting department and 4.5 labor hours from the assembly department. The profit on each new tent is $60. What is the effect of having 38 hours and 90 hours available per day in the cutting department and the assembly department, respectively?

NOTES

Chapter 6

Probability

This chapter contains examples using the calculator to illustrate:

- Finding permutations and combinations
- Calculating probabilities and expected values
- Generating random numbers
- Calculating conditional probabilities
- Applying Bayes' Formula
- Finding expected values

EXAMPLE 1 ***TEXTBOOK SECTION 6-1 & 6-2***

A group of 12 people are meeting.

(A) In how many ways can 3 be chosen for one committee and 5 be chosen for a second committee? No person can serve on both committees.

(B) How would this result be affected if there were 5 people chosen for the first committee and 3 for the second committee?

(C) How would the result be affected if there were only 11 people from which to choose?

(D) Suppose the 12 people are numbered from 1 to 12. Use the random number generator in the calculator with a seed number of 6 to pick 7 of these 12 randomly.

Solution (A): This is a problem involving the fundamental theorem of counting and combinations. Use the built-in features of your calculator to find $C_{12,3}C_{9,5}$. The result is 27,720 ways.

Solution (B): The calculation would be $C_{12,5}C_{7,3}$ if 5 people were chosen for the first committee and 3 people were chosen for the second committee. The result is the same, 27,720 ways.

Solution (C): Press [2nd] [ENTRY] to replay the last entry. Change the last entry 12 to 11 and 7 to 6. Recalculate. The result is 9,240. By reducing the number of people you start with by one, the number of ways is about one-third as many.

Solution (D): Store the seed number in the storage location RAND (see Appendix Section A-11, B-11, C-11 or D-11 of this manual). [IPart] (12 x [Rand]) will generate integers x such that $0 \le x \le 11$. Since we want integers $1 \le x \le 12$, we add 1 to this expression. So we enter [IPart] (12 x [Rand]) +1 in the calculator to get the numbers:

6	2	5	10	2	7	6	9	10
8	10	9	9	1	1	12	12	5
11	7	5	8	2	11	10	1	5

Since we cannot have repeats, the people picked are numbered

6 2 5 10 7 9 8.

EXAMPLE 2 ***TEXTBOOK SECTION 6-3***

(A) Use the random number generator to get a sample of 40 tosses of a nickel and a dime.

(B) How do the results compare to what you expected from an equally likely sample space?

Solution (A): Choose a seed number (say 3). Store this in the storage location RAND (see Appendix Section A-11, B-11, C-11 or D-11 of this manual). Suppose 0 represents no heads, 1 represents a head on the nickel and a tail on the dime, 2 represents a tail on the nickel and a head on the dime, and 3 represents two heads. We wish to generate random numbers x, where x is an integer and $0 \le x \le 3$. A method that works on all calculators is to enter [IPart] (4 x [Rand]) [ENTER]. The resulting numbers are:

0	2	2	1	0	1	0	3	1	3
1	3	1	0	0	1	3	0	1	1
2	1	2	1	3	0	0	1	1	2
0	3	2	3	2	2	3	2	3	0

Counting we see the results are: Ten 0's, twelve 1's, nine 2's, and nine 3's.

Solution (B): The expected result was 10 in each category.

EXAMPLE 3 — *TEXTBOOK SECTION 6-4*

Simulate 36 repetitions of rolling a pair of dice and recording their sum. Display a histogram of the results. The generated data and the histograms will be different for different calculators and different seed values.

Solution: Choose a seed number (say 5). Store this in the storage location RAND (see Appendix Section A-11, B-11, C-11 or D-11 of this manual).

TI-82

[5] [STO▸] [MATH] [▶] [▶] [▶] [1] <rand> [ENTER]
[MATH] [▶] [2] <iPart> [(] [10] [×] [MATH] [▶] [▶] [▶] [1] <rand> [)] [+]
[MATH] [▶] [2] <iPart> [(] [10] [×] [MATH] [▶] [▶] [▶] [1] <rand> [)]
[ENTER] [ENTER] [ENTER] ... [ENTER] [ENTER]

Repeat pressing the [ENTER] key until you have 36 numbers that are between 2 and 12 inclusive.

The resulting numbers are:

9	4	8	5	9	~~16~~	12	4	5	4
~~13~~	12	11	~~15~~	~~1~~	11	10	~~16~~	11	9
8	6	10	10	6	~~15~~	2	~~1~~	3	~~13~~
3	3	2	~~17~~	~~1~~	8	~~14~~	3	11	6
10	8	~~13~~	10	9	6	7	8		

The crossed out numbers above are discarded because they are not between 2 and 12. Store these numbers in the list L1.

Deselect or clear all functions stored in the Y= list. Set the viewing window variables to [0, 14]1 by [0, 10]1 and graph the histogram.

[2nd] [STAT PLOT] [ENTER] [ENTER] [▼] [▶] [▶] [▶] [ENTER]
[▼] [ENTER] [▼] [ENTER] [GRAPH]

TI-83

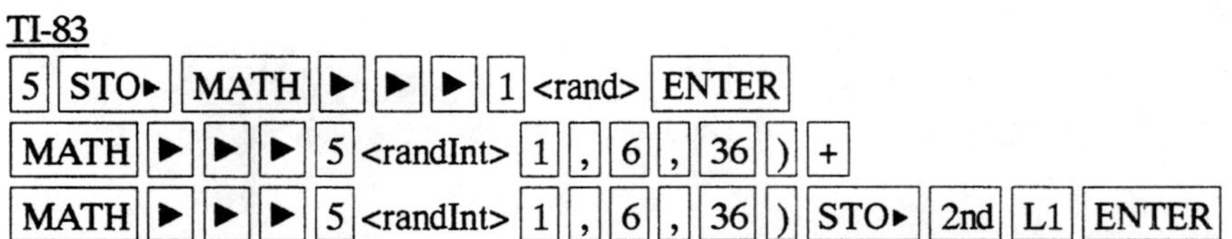

[5] [STO▸] [MATH] [▶] [▶] [▶] [1] <rand> [ENTER]
[MATH] [▶] [▶] [▶] [5] <randInt> [1] [,] [6] [,] [36] [)] [+]
[MATH] [▶] [▶] [▶] [5] <randInt> [1] [,] [6] [,] [36] [)] [STO▸] [2nd] [L1] [ENTER]

This generates 36 numbers between 1 and 6 inclusive and stores the numbers in the L1 list.

The resulting numbers are:

9	6	2	9	5	7	6	3	8	5
11	8	6	8	5	9	3	4	2	4
5	9	9	9	6	6	6	8	2	3
8	12	3	6	10	7				

Deselect or clear all functions stored in the Y= list. Set the viewing window variables to [0, 14]1 by [0, 10]1 and graph the histogram.

[2nd] [STAT PLOT] [ENTER] [ENTER] [▼] [▶] [▶] [ENTER]

[▼] [ENTER] [▼] [ENTER] [GRAPH]

<u>TI-85</u>

[5] [STO▸] [2nd] [MATH] [F2] <PROB> [F4] <rand> [ENTER]

[EXIT] [F1] <NUM> [F2] <iPart> [(] [10] [×] [EXIT] [F2] <PROB> [F4] <rand> [)] [+]

[EXIT] [F1] <NUM> [F2] <iPart> [(] [10] [×] [EXIT] [F2] <PROB> [F4] <rand> [)]

[ENTER] [ENTER] [ENTER] ... [ENTER] [ENTER]

Repeat pressing the [ENTER] key until you have 36 numbers that are between 2 and 12 inclusive.

The resulting numbers are:

9	4	8	5	9	~~16~~	12	4	5	4
~~13~~	12	11	~~15~~	~~1~~	11	10	~~16~~	11	9
8	6	10	10	6	~~15~~	2	~~1~~	3	~~13~~
3	3	2	~~17~~	~~1~~	8	~~14~~	3	11	6
10	8	~~13~~	10	9	6	7	8		

The crossed out numbers above are discarded because they are not between 2 and 12.
Store these numbers in the list xStat with 1 as the yStat values.

Deselect or clear all functions stored in the y= list.
Set the viewing window variables to [0, 14]1 by [0, 10]1 and graph the histogram.

[STAT] [F3] <DRAW> [F1] <HIST> [CLEAR]

<u>TI-86</u>

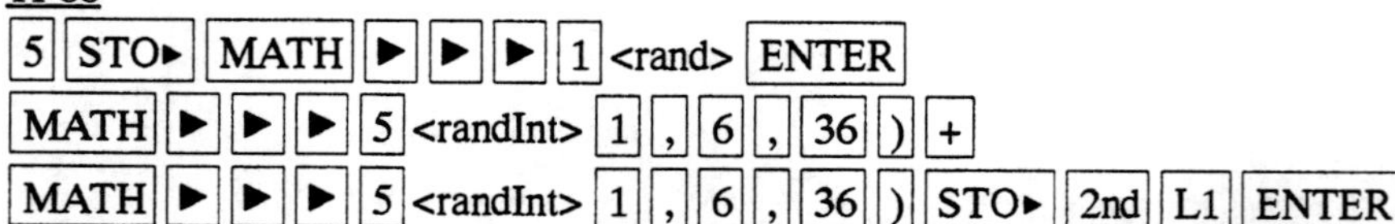

This generates 36 numbers between 1 and 6 inclusive and stores the numbers in the L1 list.

The resulting numbers are:

9	6	2	9	5	7	6	3	8	5
11	8	6	8	5	9	3	4	2	4
5	9	9	9	6	6	6	8	2	3
8	12	3	6	10	7				

Deselect or clear all functions stored in the y= list.
Store 1 as the frequency for each of the 36 pieces of data entered in the xStat list.
Set the viewing window variables to [0, 14]1 by [0, 10]1 and graph the histogram.
[2nd] [STAT] [F4] <DRAW> [F1] <HIST> [CLEAR]

EXAMPLE 3 ***TEXTBOOK SECTION 6-5***

In a study to determine employee voting patterns in a recent strike election, employees were selected at random and the following tabulation was made:

		SALARY CLASSIFICATION		
		Hourly(H)	Salary(S)	Salary+Bonus(B)
TO	Yes(Y)	394	175	20
STRIKE	No(N)	142	111	123

(A) Convert this table to a probability table.

(B) What is the probability of a randomly selected employee from this group voting to strike (Y) given that the person is paid hourly (H)?

(C) Are the events H and Y independent?

Round all answers to three decimal places.

Solution (A): Store this data as a 2×3 matrix A. Then calculate the sum of the elements. This can be done by recalling the elements using the features of the calculator, or simply adding the six numbers in the calculation mode. The result is 965. Store this answer as T using Ans→T. Then calculate $\frac{1}{T}[A]$ and store as matrix B. This is the probability matrix:

$$B = \frac{1}{T}\begin{bmatrix} 394 & 175 & 20 \\ 142 & 111 & 123 \end{bmatrix} = \begin{bmatrix} .408 & .181 & .021 \\ .147 & .115 & .127 \end{bmatrix}$$

Solution (B): We wish to calculate $P(Y|H) = \frac{B(1,1)}{B(1,1)+B(2,1)}$. Do this by recalling the elements of matrix B. Reentering the values of matrix B will cause round-off errors. In the calculation mode of the calculator enter **[B](1,1)** ÷ (**[B](1,1)** + **[B](2,1)**) where [B](1,1) represents the way to recall the element in the first row and first column of matrix B. It is shown in bold here to indicate that a single number is stored in the calculator in each location. The result is .735 .

Solution (C): We wish to see if P(Y|H) = P(H). P(Y|H) was calculated in Part (B). Calculate P(H) by calculating (**[A](1,1)** + **[A](2,1)**) ÷ T. The result is .555 . Hence the events H and Y are not independent since .555 ≠ .735 .

EXAMPLE 4 ***TEXTBOOK SECTION 6-6***

A new inexpensive skin test is devised for detecting tuberculosis. To evaluate the test before it is put into use, a medical researcher randomly selects 1,000 people. Using precise but more expensive methods already available, it is found that 8% of the 1,000 people tested have tuberculosis. Now each of the 1,000 subjects is given the new skin test and the following results are recorded: The test indicates tuberculosis in 96% of those who have it and in 2% of those who do not. Based on these results, what is the probability of a randomly chosen person having tuberculosis, given that the skin test indicates the disease?

Solution: A tree diagram is shown below that represents this situation.

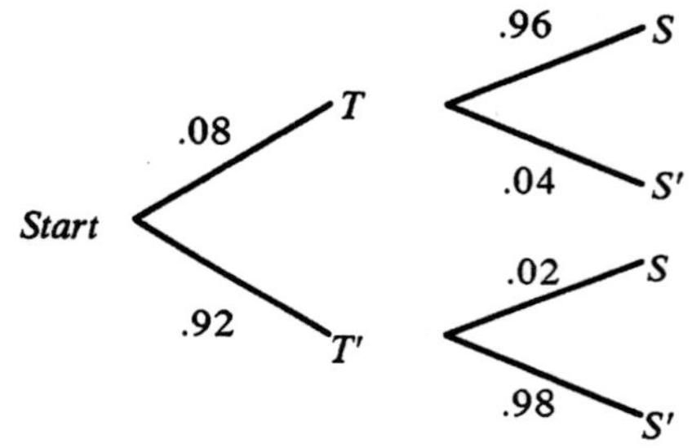

T = Tuberculosis
T' = No tuberculosis
S = Skin test indicates tuberculosis
S' = Skin test indicates no tuberculosis

We are interested in finding P(T|S). This is given by:

$$P(T|S) = \frac{P(T \cap S)}{P(T \cap S) + P(T' \cap S)} = \frac{.08 \times .96}{.08 \times .96 + .92 \times .02} = .81$$

To calculate this, enter the following into the calculator:

(.08 × .96) ÷ (.08 × .96 + .92 × .02).

EXAMPLE 5 ***TEXTBOOK SECTION 6-7***

A spinner device is numbered from 1 to 6 and each of the six numbers has a probability of occurring according to the chart below. Any winning bet of $1 will pay $5. A player always bets $1 on the 5. If the spinner lands on the 5 the player wins $4 plus gets the $1 bet back; otherwise, the $1 bet is lost. What is the expected value of the game to this player (long-run average gain or loss per game)? Use two decimal place accuracy. Is this the best number for this player to choose?

Numbers	*1*	*2*	*3*	*4*	*5*	*6*
Probability	$\frac{1}{21}$	$\frac{2}{21}$	$\frac{3}{21}$	$\frac{4}{21}$	$\frac{5}{21}$	$\frac{6}{21}$

Solution: The payoff table for this player is:

x_i	$4	-$1
p_i	$\frac{5}{21}$	$\frac{16}{21}$

To calculate the expected value enter 4×(5÷21) -1×(16÷21). The result is .19. In the long run this player will win $.19 each game on the average.

Use the replay key to calculate the other possibilities:

Number bet on	*1*	*2*	*3*	*4*	*5*	*6*
Expected value	-.76	-.52	-.29	-.05	.19	.43

The best choice for this player would be to bet on the 6 each game. In the long run this player would then win $.43 each game on the average.

Exercise Set 6

1. Consider a group of 18 people:

   (A) In how many ways can 4 be chosen for one committee and 3 be chosen for a second committee? A person can serve on both committees.

   (B) How would the result from Part (A) be affected if there were 5 people chosen for the first committee and 3 for the second committee?

   (C) How would the result from Part (A) be affected if there were only 15 people?

   (D) How would the result from Part (B) be affected if there were only 15 people?

2. From a group of 20 books, in how many ways can 5 be chosen for one bookrack and 10 be chosen for a second bookrack and 3 be chosen for a third bookrack. How would this result be affected if there were 25 books to start with?

3. Twenty-five paintings are available for purchase at an art show. The paintings are numbered from 0 to 24. The artist wishes to choose five randomly to sell for charity. What are the painting numbers that he should choose?

4. A group of 30 people are meeting. We want to randomly choose 10 of these people to work on a specific task. If the people are numbered from 1 to 30, who should be chosen?

5. Repeat Example 3 above if a third category, Undecided (U), was added to the vote a person would cast, where the Undecided (U) category had 102 Hourly, 87 Salary, and 187 Salary+Bonus votes.

6. Repeat Example 1 if the company did not have a Salary+Bonus category but did have the third category of Undecided (U) from Problem 1.

7. A company's brand (X) has 22% of the market. A market research firm finds that if a person currently uses brand X, the probability is .61 that he or she will buy it the next time (represented by X_c). On the other hand, if a person currently does not use X (represented by X_c'), the probability is .53 that he or she will switch to X the next time (represented by X_n).

   (A) Draw a tree diagram representing this situation.

   (B) What is the probability that a person is presently using brand X if he/she purchases brand X on the next purchase?

8. Repeat Problem 3 if the company only had 15% of the market.

9. A gaming device is numbered from 1 to 10 and each of the ten numbers has a probability of occurring according to the chart below. If the device lands on the chosen number the player wins $8 and gets the bet back; otherwise, the $2 bet is lost. Suppose a player always bets $2 on the 5.

Numbers	*1*	*2*	*3*	*4*	*5*	*6*	*7*	*8*	*9*	*10*
Probability	.05	.05	.06	.06	.21	.21	.08	.08	.10	.10

   (A) What is the expected value of the game to this player (long-run average gain or loss per game)? Use two decimal place accuracy.

   (B) Is this the best number for this player to choose?

10. Repeat Problem 9 if the bet is $4 and the player wins $5 and gets the bet back.

Chapter 7

Data Description and Probability Distributions

This chapter contains examples using the calculator to illustrate:

- Graphing quantitative data
- Calculating measures of central tendency
- Calculating measures of dispersion
- Calculating the probability of x successes in n Bernoulli Trials

EXAMPLE 1 ***TEXTBOOK SECTION 7-1***

(A) Create a bar graph for the data shown in the table at the right.

(B) Create a broken-line graph for the data.

US. Public Debt

Year	Billions of Dollars
1950	$ 256.1
1960	284.1
1970	370.1
1980	907.7
1990	3233.3

Solution (A): Assign a number to each year and express the debt as a whole number as shown below:

Year	1950	1960	1970	1980	1990
x	1	2	3	4	5
y	2561	2841	3701	9077	32333

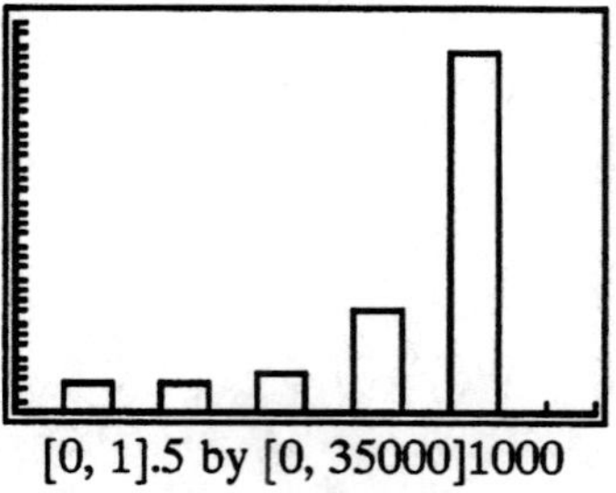

[0, 1].5 by [0, 35000]1000

Clear the functions from the function list and any drawings (see Appendix Section A-6, B-6, C-6 or D-6 of this manual). Clear the statistical registers in the calculator (see Appendix Section A-17, B-17, C-17 or D-17 of this manual).

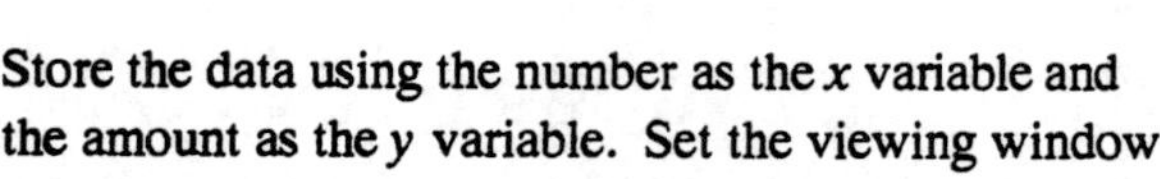

Store the data using the number as the x variable and the amount as the y variable. Set the viewing window variables at [0, 6].5 by [0, 35000]1000. The bars will be .5 in width and there will be a gap of .5 between bars because an Xscl of .5 will graph data as $0 \le x < .5$, $.5 \le x < 1$, $1 \le x < 1.5$, etc. and there is no data for $.5 \le x < 1$, $1.5 \le x < 2$, etc. The center of the bars is not at the x value, but is at the center of the bar, at $x+.25$

Solution (B): Sort the data, if it is not already sorted. Clear the drawings (see Appendix Section A-6, B-6 or C-6 of this manual). Now use the xy-line option on the statistical draw menu to get the graph at the right. See Appendix Section A-16, B-16, or C-16 Example 2(E) of this manual.

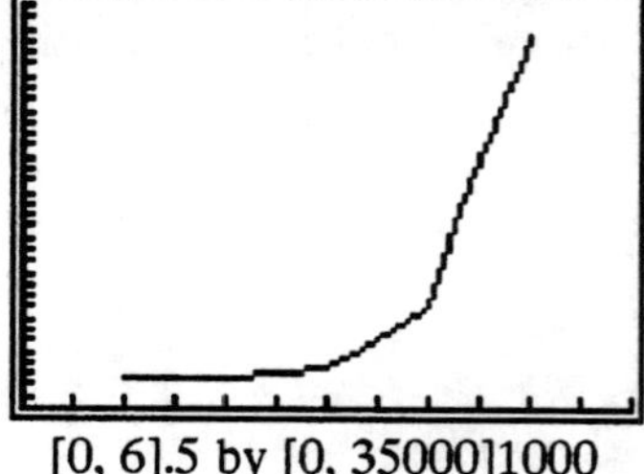

[0, 6].5 by [0, 35000]1000

The line segments are drawn from (1, 2561) to (2, 2841) to (3, 3701) to (4, 9077) to (5, 32333) which are the top left corners of the bars shown in Solution (A) above.

EXAMPLE 2 ***TEXTBOOK SECTION 7-1***

The following table shows price-earnings ratios of common stocks chosen at random from the New York Stock Exchange:

Price-Earnings (PE) Ratios

7	11	6	6	10	6	31	28	13	19
6	18	9	7	5	5	9	8	10	6
10	3	4	6	7	9	9	19	7	9
17	33	17	12	7	5	7	10	7	9
18	17	4	6	11	13	7	6	10	7

Construct a histogram using 0 as the lower limit, 40 as the upper limit, and class intervals of 5. Use the arrow keys on your calculator to move the cursor and explore the graph. Where do you think the "center" of the data is?

Solution: (See Appendix Section A-16, B-16, C-16 or D-16 of this manual on how to enter data and get a histogram for your calculator.) Set the viewing window variables to [0,40]5 by [-5,30]5. The result is shown to the right. Remember that the Xscl is the class width.

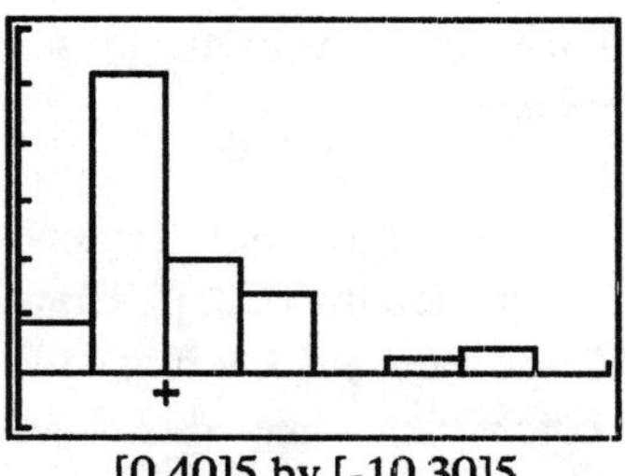
[0,40]5 by [-10,30]5

The "center" of the data is usually thought of as being the balance point or the mean. The cursor shown on the graph to the right is probably approximately where you placed your cursor. (Now you see the reason for setting the limits for the *y* axis to start below zero. This gives room on the graph to see the cursor and the coordinates of its location. The coordinates are not shown on the graph above because each calculator model will have different coordinates.)

EXAMPLE 3 ***TEXTBOOK SECTION 7-2***

Given the data below:

(A) Create a frequency polygon for this data.

(B) Create a relative cumulative frequency ogive for this data.

Class Interval	Midpoint x	Frequency y	Relative Cumulative Frequency
299.5-349.5	324.5	1	.01
349.5-399.5	374.5	2	.03
399.5-449.5	424.5	5	.08
449.5-499.5	474.5	10	.18
499.5-549.5	524.5	21	.39
549.5-599.5	574.5	20	.59
599.5-649.5	624.5	19	.78
649.5-699.5	674.5	11	.89
699.5-749.5	724.5	7	.96
749.5-799.5	774.5	4	1.00

Solution (A): Add a line to the table both at the beginning and at the end.

249.5-299.5	274.5	0	.00
799.5-849.5	824.5	0	1.00

Store the data in the calculator so that x=midpoint and y=frequency.

Sort the data on x. Set the viewing window variables at [250, 850]50 by [0, 25]1. Draw the xy-line. The TI-82 will show the dots if you draw the scatter graph first before the xy-line. (See Appendix Section A-16, B-16, C-16 or D-16 of this manual.)

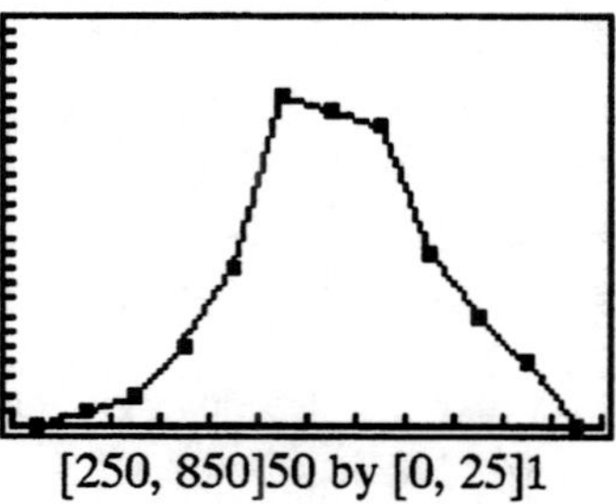

[250, 850]50 by [0, 25]1

Solution (B): Add a line at the beginning of the original table:

249.5-299.5	274.5	0	.00

The calculator needs to have a whole number for the y variable in order to draw the xy-line. Change the relative cumulative frequency to whole numbers by multiplying each by a number, say 100. Store the data in the calculator so that x=upper class boundary and y=whole number representing the relative cumulative frequency. Set the viewing window variables at [250, 850]50 by [0, 150]10. Draw the xy-line. The TI-82, TI-83, or TI-86 will show the dots if you draw the scatter graph first before the xy-line.

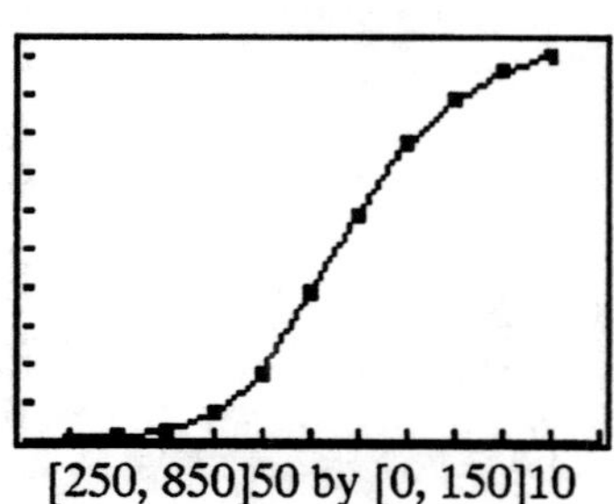

[250, 850]50 by [0, 150]10

EXAMPLE 4 ***TEXTBOOK SECTION 7-3***

(A) Find the median and mode of the data in Example 2 above.

(B) Find the median and mode of the data in Example 3 above.

Solution (A): Clear the statistical registers in the calculator and enter the data as was done in Example 2 above.

TI-82, TI-83, and TI-86

These calculator models have a built-in function that will find the median for you. See Appendix Section A-16, B-16 or D-16 of this manual. The median function is on the 1-Var Stats list. (Use the down arrow to see the last part of the list.) The median is 9 and the mode is 7.

TI-85

Use the calculator to sort the data on the x variable. (See Appendix Section C-16 of this manual.) There are 50 pieces of data so the median value is between the 25th and 26th value. Use the arrow keys while in the edit mode to examine the data to find the values in the 25th and 26th position. Take the average of these two values. The median is 9.

The mode is 7.

Solution (B): Clear the statistical registers in the calculator and enter the data as was done in Example 3(A) above.

TI-82, TI-83 and TI-86

These calculator models have a built-in function that will find the median for you. See Appendix Section A-16, B-16 or D-16 of this manual. The median function is on the 1-Var Stats list. (Use the down arrow to see the last part of the list.) The median is listed is 574.5. This is the midpoint of the class of the 50.5th value. However, if we want the number that is 11.5/20ths into the class, we must use the method shown for the TI-85 calculators below.

TI-85

Use the calculator to sort the data on the x variable. (See Appendix Section C-16 of this manual.) There are 100 pieces of data so the median value is between the 50th and 51st value. Use the arrow keys in the edit mode and count down the frequencies of the y variable) until you find the class having the 50th piece of data. This is the sixth class. Since there are 39 numbers before this class, we need to find the numbers that is 11.5th position into the sixth class. We find the median by taking the fraction of the class width and adding it on to the lower boundary of the class: $\frac{11.5}{\text{number in class}} \times$ (class width) + (lower class boundary) = $\frac{11.5}{20} \times 50 + 549.5 = 578.25$. Hence the median is 578.25.

The modal class has midpoint 524.5. This can be observed in the tabled data or by inspecting the histogram with range [249.5, 824.5]50 by [-5, 25]1.

EXAMPLE 5 ***TEXTBOOK SECTION 7-3 & 7-4***

Find the mean and standard deviation to two decimal places for the data in Example 2 above.

Solution: Store the data in the calculator with x=the data value and y=the frequency (this is 1 if entering each piece of data separately). (See Appendix Section A-16, B-16, C-16 or D-16 of this manual on how to get the one-variable statistics on your calculator.)

The results are: $\bar{x} = 10.52$ and $s = 6.63$. Since this is a sample of data, we use the sample standard deviation and not population standard deviation.

EXAMPLE 6 ***TEXTBOOK SECTION 7-3 & 7-4***

Find the mean and standard deviation to two decimal places for the data in Example 3 above.

Solution: Store the data using the midpoint of the class as the x variable and the frequency as the y variable as was done in Example 3(A) above. (See Appendix Section A-16, B-16, C-16 or D-16 of this manual on how to get the one-variable statistics on your calculator.) The results are: $\bar{x} = 579$ and $s = 94.31$. Since this is a sample of data, we use the sample standard deviation and not population standard deviation.

EXAMPLE 7 ***TEXTBOOK SECTION 7-5***

The probability of recovering after a particular type of operation is .95. Find the probability that at least 5 of the next 50 patients having this operation will not recover.

Solution: At least 5 not recovering is the same as at most 45 will recover. So we want to find the probability of at least five patients not recovering is equivalent to less than 45 patients recovering.

$$\begin{aligned} P(X \le 45) &= P(1) + P(2) + \ldots + P(45) \\ &= 1 - P(X > 45) \\ &= 1 - P(X \ge 46) \\ &= 1 - [P(46) + P(47) + P(48) + P(49) + P(50)] \end{aligned}$$

We need to find P(46)+ P(47)+ P(48)+ P(49)+ P(50) and subtract the result from 1.

This is a binomial distribution situation with $n = 50, p = .95$, and $q = .05$ when a recovery is considered a success. The probability of x people recovering is given by: $P(x) = C_{50,x}\,(.95)^x(.05)^{50-x}$. Store this expression in the calculator as: (50 **nCr** X) (.95 ^ X) .05 ^ (50 - X) where nCr is found on the probability menu (see Appendix

Section A-11, B-11, C-11 or D-11 of this manual). Store 46 for X and evaluate the expression. Store the result in memory M. Store 47 for X and reevaluate the expression. Add the result to the contents of memory M and store again as memory M. Continue in this manner until the sum of the probabilities is in memory M.

(50 nCr X) (.95 ^ X) .05 ^ (50 - X)
46 → X
Recall the expression and evaluate.
Ans → M Result is .136

47 → X
Recall the expression and evaluate. Result is .220
Ans + M → M Result is .356

48 → X
Recall the expression and evaluate. Result is .261
Ans + M → M Result is .617

49 → X
Recall the expression and evaluate. Result is .202
Ans + M → M Result is .819

50 → X
Recall the expression and evaluate. Result is .077
Ans + M → M Result is .896

Find 1 - M. The result is .104. Thus the probability that at least 5 people out of 50 do not survive the operation is .104 or a little more than 1 chance in 10.

EXAMPLE 8 *TEXTBOOK SECTION 7-6*

Given the normal probability distribution having mean 15.4 and standard deviation 2.3:

(A) Graph using [6, 24]1 by [0, .3].1.

(B) Find the value of c so that $P(9.2 \le x \le c) = .8534$.

(C) Conjecture, then check by graphing, the location and shape of a normal probability distribution having mean 17 and standard deviation 1.5.

Solution (A): The formula for this situation is: $f(x) = \frac{1}{2.3\sqrt{2\pi}} e^{-\frac{(x - 15.4)^2}{2(2.3^2)}}$.

Store this in the calculator as:

1 ÷ (2.3 x √ (2 x π)) x [e^] (- ((X - 15.4) ^ 2) ÷ (2 x 2.3 ^ 2)).

(Recall that [e^] is found using [2nd] [LN].)

Set the viewing window variables to [6, 24]1 by [0, .3].1. Graph.

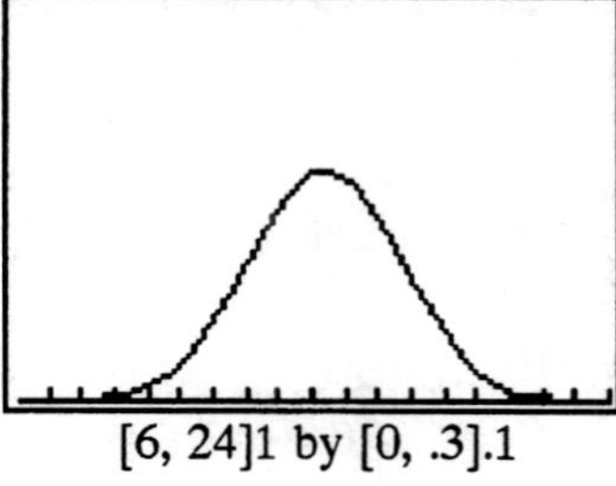

[6, 24]1 by [0, .3].1

Solution (B): The z value for x=9.2 is $z = \frac{9.2-15.4}{2.3} = -2.70$. From the table we find the area under the curve for z between -2.70 and 0 is .4965. Hence we know that the x value we want is on the right side of the mean which is 15.4. .8534-.4965=.3569 is the area on the right side of x=15.4. This corresponds to a z of .98.

The c value is found by solving $.98 = \frac{c - 15.4}{2.3}$. $c = 17.65$.

Solution (C): The graph will be shifted to the right 1.6 units and will be taller and skinnier since the standard deviation is smaller.

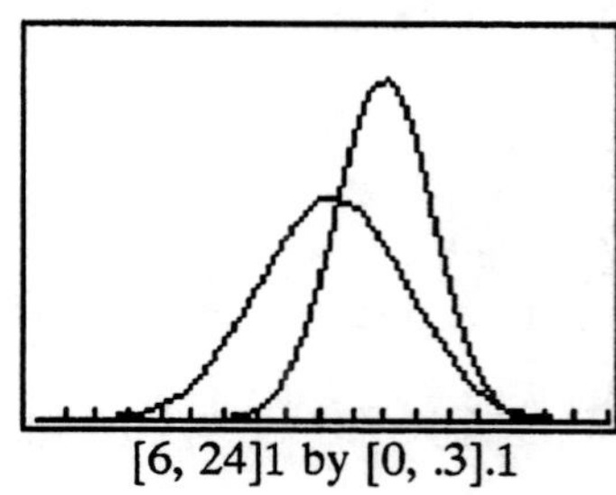

[6, 24]1 by [0, .3].1

Exercise Set 7

1. The following table shows scores of students on a 35 question true/false test.

   Scores on True/False test

27	31	16	26	10	16	31	28	13	19	16	28	29	17
25	25	29	18	10	26	20	13	24	26	27	19	29	19
17	19	27	33	17	12	27	25	17	10	7	19	18	27
34	26	11	13	17	26	10	27						

   Construct a histogram using 0 as the lower limit, 35 as the upper limit, and class intervals of 5.

   Use the arrow keys on your calculator to move the cursor and explore the graph. Where do you think the "center" of the data is?

2. Suppose the first column of scores in Problem 7 were all perfect papers with scores of 35. Repeat Problem 1 and discuss how this change of data effects the histogram.

3. Find the mean and standard deviation of the sample data in Problem 1.

4. Find the mean and standard deviation of the sample data in Problem 2.

5. Repeat Example 7 but find the probability that 3, 4 or 5 will not recover.

6. Repeat Example 7 but change the probability of recovering to .97.

NOTES

NOTES

Chapter 8

Games and Decisions

This chapter contains examples using the calculator to illustrate:

- Finding the expected value of a game
- Finding the solution to a 2×2 nonstrictly determined matrix game

EXAMPLE 1 ***TEXTBOOK SECTION 8-2***

Find the expected value of the two-finger Morra game for R if strategies

$P = [.248 \quad .752]$ and $Q = \begin{bmatrix} .381 \\ .619 \end{bmatrix}$ are used by R and C, respectively, and the game matrix is $M = \begin{bmatrix} a & b \\ c & d \end{bmatrix} = \begin{bmatrix} 2 & -3 \\ -3 & 4 \end{bmatrix}$.

Solution: The expected value of the game for R is given by PMQ. Store the matrix P as a 1x2 matrix A, and M as a 2x2 matrix B in the calculator. Store matrix Q as a 2x1 matrix C in the calculator. Multiply the matrices $A*B*C$ to find the product PMQ. (See Appendix Section A-12, B-12, or C-12 of this manual.) The result is .731. This means that in the long run, with R and C using the indicated strategies, R will win \$.73 per game and C will lose \$.73 per game, on the average.

EXAMPLE 2 ***TEXTBOOK SECTION 8-2***

(A) Solve the two-finger Morra game of Example 1 above.

(B) Determine if the game is fair.

Solution (A): We need to find the optimal strategy for R and C. The optimal strategy for R is given by $P^* = \begin{bmatrix} \frac{d-c}{D} & \frac{a-b}{D} \end{bmatrix}$ and the optimal for C is given by $Q^* = \begin{bmatrix} \frac{d-b}{D} \\ \frac{a-c}{D} \end{bmatrix}$ where $D = (a + d) - (b + c)$. Calculate D and store in memory location D. The result is 12. Now calculate the other quantities as (4-(-3))÷D, (2-(-3))÷D, etc. The result is $P^* = [.583 \quad .417]$ and $Q^* = \begin{bmatrix} .583 \\ .417 \end{bmatrix}$.

Solution (B): We need to check to see if the expected value $\frac{\det M}{D}$ is 0. Calculate determinant of M using the calculator. (See Appendix Section A-12, B-12, or C-12 of this manual.) The result is -1. The expected value for R is $-\frac{1}{12}$. Since the expected value of the game is not zero, this is not a fair game.

Exercise Set 8

1. Find the expected value of the two-finger Morra game for R if strategies

   $P = [.187 \quad .813]$ and $Q = \begin{bmatrix} .381 \\ .619 \end{bmatrix}$ are used by R and C, respectively, and the game matrix is $M = \begin{bmatrix} a & b \\ c & d \end{bmatrix} = \begin{bmatrix} 2.5 & -3.2 \\ -3.4 & 4.8 \end{bmatrix}$.

2. Find the expected value of the two-finger Morra game for R if strategies

   $P = [.683 \quad .317]$ and $Q = \begin{bmatrix} .381 \\ .619 \end{bmatrix}$ are used by R and C, respectively, and the game matrix is $M = \begin{bmatrix} a & b \\ c & d \end{bmatrix} = \begin{bmatrix} 3.2 & -2.7 \\ -2.8 & 6.2 \end{bmatrix}$.

3. Find the optimal strategies for the two-finger Morra game of Problem 1 and determine if the game is fair.

4. Find the optimal strategies for the two-finger Morra game of Problem 2 and determine if the game is fair.

Chapter 9

Markov Chains

This chapter contains examples using the calculator to illustrate:

- Finding next state and stationary state matrices in Markov chains
- Finding powers of the transition matrix
- Determining if a transition matrix is regular
- Finding a limiting matrix of a transition matrix

EXAMPLE 1 ***TEXTBOOK SECTION 9-1, 9-2, and 9-3***

A given plant species has red, pink, or white flowers according to the genotypes RR, RW, and WW, respectively. If each of these genotypes is crossed with a pink-flowering plant (genotype RW), then the transition matrix P is:

$$P = \begin{array}{c} \\ \textbf{This} \\ \textbf{generation} \end{array} \begin{array}{c} \\ \text{Red} \\ \text{Pink} \\ \text{White} \end{array} \begin{array}{c} \textbf{Next generation} \\ \begin{array}{ccc} \text{Red} & \text{Pink} & \text{White} \end{array} \\ \left[\begin{array}{ccc} .50 & .50 & .00 \\ .25 & .50 & .25 \\ .00 & .50 & .50 \end{array}\right] \end{array}$$

Assuming the plants of each generation are crossed only with pink plants to produce the next generation:

(A) Find, to two decimal places, the percentage of each color plant after four generations and after 10 generations if the initial generation is all red plants.

(B) Show that, regardless of the makeup of the first generation, the genotype composition will eventually stabilize at 25% red, 50% pink, and 25% white.

(C) Repeat Parts (A) and (B) if the initial generation is all white plants.

(D) Conjecture, then check, what the fourth generation, tenth generation, and stationary matrix will be if the initial matrix is $S_0 = [0 \quad 1 \quad 0]$.

(E) Is P a regular matrix?

(F) Find the limiting matrix $\bar{P}$.

Solution (A): We need to find the fifth state matrix with initial matrix of $S_0 = [1 \quad 0 \quad 0]$. Store the transition matrix P in the calculator. The fourth state matrix is found by calculating $S_0P^4 = [1 \quad 0 \quad 0]\begin{bmatrix} .50 & .50 & .00 \\ .25 & .50 & .25 \\ .00 & .50 & .50 \end{bmatrix}^4$. To do this, store the 3×3 matrix $P = \begin{bmatrix} .50 & .50 & .00 \\ .25 & .50 & .25 \\ .00 & .50 & .50 \end{bmatrix}$ as matrix A in the calculator. Also, store the 1×3 matrix $S_0 = [1 \quad 0 \quad 0]$ as matrix B in the calculator. (See Appendix Section A-12, B-12, C-12, or D-12 of this manual.) Find the fourth power of A and store it as matrix C. The result is $\begin{bmatrix} .28125 & .50000 & .21875 \\ .25000 & .50000 & .25000 \\ .21875 & .50000 & .28125 \end{bmatrix}$.

Now multiply the initial state matrix by matrix C. The result is [.28125 .50000 .21875].

An alternative method would be to find BA^4 directly using matrix multiplication without finding A^4 first.

We also need to find $S_0P^{10} = BA^{10}$. The result is [. 25049 .50000 .24951]. We will see that this is very close to the stationary matrix.

Solution (B): We need to find the stationary matrix. This matrix is found by solving the matrix equation:

$$[m \quad n \quad p]\begin{bmatrix} .50 & .50 & .00 \\ .25 & .50 & .25 \\ .00 & .50 & .50 \end{bmatrix} = [m \quad n \quad p].$$ This is the same as

$$[m \;\; n \;\; p]\begin{bmatrix} .50 & .50 & .00 \\ .25 & .50 & .25 \\ .00 & .50 & .50 \end{bmatrix} = [m \;\; n \;\; p]\begin{bmatrix} 1 & 0 & 0 \\ 0 & 1 & 0 \\ 0 & 0 & 1 \end{bmatrix}$$

$$[m \;\; n \;\; p]\left\{\begin{bmatrix} .50 & .50 & .00 \\ .25 & .50 & .25 \\ .00 & .50 & .50 \end{bmatrix} - \begin{bmatrix} 1 & 0 & 0 \\ 0 & 1 & 0 \\ 0 & 0 & 1 \end{bmatrix}\right\} = [\,0 \;\; 0 \;\; 0\,]$$

$$[m \;\; n \;\; p]\begin{bmatrix} -.50 & .50 & .00 \\ .25 & -.50 & .25 \\ .00 & .50 & -.50 \end{bmatrix} = [\,0 \;\; 0 \;\; 0\,].$$ This result is equivalent to

$$\begin{aligned} -.50m + .25n \quad\quad &= 0 \\ .50m - .50n + .50p &= 0. \\ .25n - .50p &= 0 \end{aligned}$$

This is a dependent system.

Only two of these equations can be used. The third equation to use in finding the stationary matrix is $m+n+p=1$. Hence we need

to solve the matrix equation $\begin{bmatrix} -.50 & .25 & .00 \\ .50 & -.50 & .50 \\ 1.00 & 1.00 & 1.00 \end{bmatrix}\begin{bmatrix} m \\ n \\ p \end{bmatrix} = \begin{bmatrix} 0 \\ 0 \\ 1 \end{bmatrix}$. Enter the 3×3 coefficient matrix into the calculator as matrix A and find its inverse. This result is

$\begin{bmatrix} -2.00 & -.50 & .25 \\ .00 & -1.00 & .50 \\ 2.00 & 1.50 & .25 \end{bmatrix}$. Store this as matrix C. Store the constant 3×1 matrix as matrix B in the calculator. Multiply the inverse matrix by the constant matrix, CB. The result is $\begin{bmatrix} .25 \\ .50 \\ .25 \end{bmatrix}$. The stationary matrix is then [.25 .50 .25] to two decimal place accuracy. The expected genotype composition will eventually stabilize at 25% red, 50% pink, and 25% white.

Solution (C): We need to find the product $S_0P^4 = BA^4 =$

$$[\,0 \;\; 0 \;\; 1\,]\begin{bmatrix} .50 & .50 & .00 \\ .25 & .50 & .25 \\ .00 & .50 & .50 \end{bmatrix}^4$$

. Store the 1×3 matrix S_0 as matrix B in the calculator. Store the 3×3 matrix P as matrix A in the calculator. Find the matrix product BA^4. The result is [.21875 .50000 .28125].

The tenth state matrix is [.24951 .50000 .25049].

The steady-state matrix is the same as found in Part (B) above.

Solution (D): A good conjecture would be that the fourth state matrix would be close to [.24 .52 .24], that the tenth state matrix would be close to the stationary matrix, and that the stationary matrix is the same as in Parts (B) and (C).

As it turns out, $S_0P^4 = BA^4 = [0\ \ 1\ \ 0] \begin{bmatrix} .50 & .50 & .00 \\ .25 & .50 & .25 \\ .00 & .50 & .50 \end{bmatrix}^4 = [\ .25\ .50\ .25\ ]$ = stationary matrix.

Solution (E): P is a regular matrix since A^2 has all positive elements.

Solution (F): Find successive powers of P to find $\bar{P}$. Do this on the calculator using matrix A since this is where we stored the elements of matrix P.

$\bar{P} = \begin{bmatrix} .25 & .50 & .25 \\ .25 & .50 & .25 \\ .25 & .50 & .25 \end{bmatrix}$. We need to find A^{34} (A^{40} on the TI-85) to see the limiting matrix. An easy way to do this is to recall matrix A. Then enter [Ans] × [A]. Pressing the [ENTER] key repeatedly will continue to perform this multiplication. Count the number of times you push the [ENTER] key.

Exercise Set 9

1.. A company's brand X has 22% of the market. A market research firm finds that if a person uses brand X now, represented by X_0, the probability is .61 that he or she will buy it the next time, represented by X_1. On the other hand, if a person does not use brand X now (represented by X_0'), the probability is .53 that he or she will switch to brand X the next time, represented by X_1'.

(A) Draw a probability tree diagram representing this situation.

(B) What is the probability that a person is currently using brand X, represented by X_0, if he/she purchases brand X on the next purchase, represented by X_1?

2. Repeat Problem 1 if the company only had 15% of the market.

3. Refer to Problem 1 above. (A) What share of the market will the company have after 5 purchases? (B) What share of the market will the company have in the long run?

4. Repeat Problem 1 if in Problem 1 the probability is .75 that a person currently using brand X, represented by X_0, will purchase it the next time.

5. Repeat Problem 1 if the company introduced a new product Y so that the initial matrix is [.35 .21 .44] and the transition matrix is:

	X_1	Y_1	*neither* X_1 *nor* Y_1
X_0	.58	.29	.13
Y_0	.35	.45	.20
neither X_0 *nor* Y_0	.13	.17	.70

6. Repeat Problem 5 where $P(X_1|X_0) = .63$, $P(Y_1|X_0) = .21$, $P(neither|X_0) = .16$ with the remaining probabilities the same. In other words, the first row of the matrix in Problem 9 is .63, .21, .16.

NOTES

NOTES

Appendix A

TI-82 Operations

A-1 Operations

```
MEMORY
1: Check RAM...
2: Delete...
3: Reset...
```

```
RESET MEMORY
1: No
2: Reset
```

Press [ON] to turn on the calculator.

Press [2nd] [+] to get the **MEMORY** screen (shown at the right).

Choose 3:Reset... by pressing [3] or use the down-arrow key [▼] to highlight 3:Reset... and press [ENTER].

The display shows the **RESET MEMORY** .

Choose 2: Reset or use the down-arrow key to highlight 2:Reset... and press [ENTER]. You may need to adjust the contrast (directions follow).

`Mem cleared` should appear on the screen.

If it doesn't, press [2nd] and then hold the [▲] (to increase the contrast) or [2nd] [▼] (to decrease the contrast) until you see it on the screen.

Press [CLEAR] to clear the screen.

Press [2nd] [OFF] to turn the calculator off.

The [2nd] Key

Press this key to access the operation listed on the face of the calculator above and to the left of a key. The operations above the key are in a light blue color on the face of the calculator. Note that a flashing up arrow [↑] appears on the screen after [2nd] key is pressed.

> In this manual, the functions on the face of the calculator above a key will be referred to in square boxes just as if the function was printed on the key cap. For example, [ANS] is the function above the [(–)] key.

The [ALPHA] and [A-LOCK] Keys

The operation above and to the right of a key is accessed by first pressing [ALPHA] or [A-LOCK]. A flashing [A] is displayed as the cursor on the screen after the [ALPHA] key is pressed. [A-LOCK] locks the calculator into alpha mode until the [ALPHA] is pressed.

Home Screen

Calculations are done and commands are entered on the Home Screen. You can always get to this screen (aborting any calculations in progress) by pressing

QUIT
[2nd] [MODE]. From here on, this will be referred to as [2nd] [QUIT] in this appendix.

The [MODE] Key

When you press [MODE] you see the screen shown below. The highlighted items are active.

To change an item, select it using the arrow keys and press [ENTER] to activate the selection.

Type of notation for display of numbers.	Normal Sci Eng
Number of decimal places displayed.	Float 0123456789
Type of angle measure.	Radian Degree
Function, parametric, polar or sequence graphing.	Func Par Pol Seq
Connected/dot plotted points on graphs.	Connected Dot
Graphs functions separately or all at once.	Sequential Simul
Allows a full screen or split screen to be used.	FullScreen Split

Menus

The TI-82 Graphics calculator uses menus for selection of specific functions. The items on the menus are identified by numbers followed by a colon. There are two ways to choose menu items:

1. Press the number corresponding to the menu item.
2. Use the arrow keys to highlight the selection and then pressing [ENTER].

In this manual the menu items will be referred to using the key to be pressed followed by the meaning of the menu. For example, on the ZOOM menu, [1] <ZBox> refers to the first menu item ZBox.

Editing

It is easy to correct errors when entering data into the calculator by using the arrow keys, [DEL], and [INS] keys.

[◄] or [►]	Moves the cursor to the left or right one position.
[▲]	Moves the cursor up one line.
[▼]	Moves the cursor down one line.
[DEL]	Deletes one or more characters at the cursor position.
[2nd] [INS]	Inserts one or more characters at the cursor position.
[2nd] [ENTRY]	Replays the last executed line of input.

A-2 Calculating

EXAMPLE 1: Calculate $|-4^2-\sqrt{34}|$

Enter numbers and functions in the calculator in the same order as you would read an expression. In the examples that follow, do not press [ENTER] unless specifically instructed to do so. Enter all the keystrokes without pressing the ENTER key until [ENTER] is listed in the example.

The subtraction sign and the negative sign are different.

> The **subtraction sign** is a dark blue key on the lower right side of the calculator. This is used when subtracting one quantity from another.
>
> The **negative sign** (-) is a light gray key in the bottom row of keys on the calculator. This is used when entering a negative quantity.

Solution: [2nd] [QUIT] [2nd] [ABS] [(] [(-)] [4] [x^2] [-] [2nd] [√] [34] [)] [ENTER]. The result displayed is 21.83095189.

A-3 Plotting Points and Drawing Line Segments

EXAMPLE 1: (A) Plot the points (-3, 6) and (1,5). (B) Draw the line segment having these points as the endpoints.

Solution (A):

Keystrokes	Explanation
[Y=] [CLEAR] [▼] [CLEAR] etc.	Clear all functions from the function list.
[2nd] [QUIT]	Set the viewing window dimensions.
[ZOOM] [6] <ZStandard> [CLEAR]	[ZOOM] [6] <ZStandard> automatically sets the dimensions to Xmin=-10, Xmax=10, Xscl=1, Ymin=-10, Ymax=10, and Yscl=1. This will be denoted as [-10, 10]1 by [-10, 10]1 in this manual. Press [CLEAR] to get back to the home screen.
[2nd] [DRAW] [1] <ClrDraw> [ENTER]	Get the [DRAW] menu and clear the drawings from the calculator screen.
[2nd] [DRAW] [▶] [1] <PT-On(> [(-)] [3] [,] [6] [)] [ENTER]	Get the [DRAW] menu again. Choose the point on option and enter the coordinates of the point with a comma separating them. The point will be plotted.
[2nd] [QUIT]	
[2nd] [DRAW] [▶] [1] <PT-On(> [1] [,] [5] [)] [ENTER]	Enter the coordinates of the second point and plot.

Solution (B):

Keystrokes	Explanation
[2nd] [QUIT]	Return to the Home Screen. Get the [DRAW] menu and choose the line option.
[2nd] [DRAW] [2] <Line(> [(-)] [3] [,] [6] [,]	Enter the coordinates of the points separated by commas.
[1] [,] [5] [)] [ENTER]	The line segment will be drawn between the two points.

A-4 Storing and Evaluating an Expression, Lists and Tables

EXAMPLE 1: Store the expression $\frac{-b+\sqrt{b^2-4ac}}{2a}$ and evaluate it to solve $3x^2+4x-5=0$.

Solution:

Keystrokes	Explanation
[Y=] [CLEAR] [(] [(-)] [ALPHA] [B] [+] [2nd] [√]	You will see a list of function names Y1, Y2, Y3, etc. We can store ten different functions or algebraic expressions in this list.
[(] [ALPHA] [B] [x^2] [-] [4] [ALPHA] [A] [×] [ALPHA] [C] [)] [)] [÷] [(] [2] [ALPHA] [A] [)]	Then we can recall them and evaluate or graph them. We will use Y1. Press these keys to enter the expression. (-B+√(B²-4AC))/(2A)
[ENTER] [2nd] [QUIT]	Get back to the home screen.
[3] [STO▸] [ALPHA] [A] [ENTER] [4] [STO▸] [ALPHA] [B] [ENTER] [(-)] [5] [STO▸] [ALPHA] [C] [ENTER]	Store 3 as A, 4 as B and -5 as C.
[2nd] [Y-VARS] [1] <Function> [1] <Y1> [ENTER]	Recall the expression and evaluate. The result is .79 to two decimal places. So one solution to this quadratic equation is x=.79.
[Y=] [▶] ... [▶] [-] [ENTER] etc.	Get the function list. Now change the addition sign to a subtraction sign and evaluate again. The other solution is x=-2.12.

> Throughout this manual, when using menus, the number of the desired selection will be shown followed by what the selection is. Thus, [Y-VARS] [1] <Function> means to pick the first selection (that is, Function) off the Y-VARS menu.

EXAMPLE 2: Given $f(x) = \sqrt[3]{x^4+2x}$ and $g(x) = 4^{2x-90}$, find $f(x) + 3\sqrt{g(x)}$ at $x = 45.2$.

Solution:

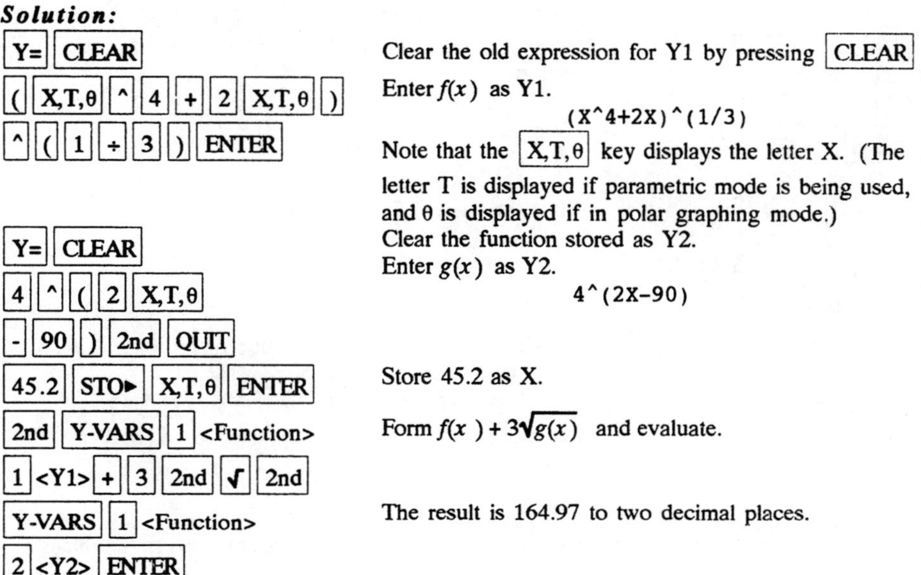

Keystrokes	Explanation
[Y=] [CLEAR]	Clear the old expression for Y1 by pressing [CLEAR]
[(] [X,T,θ] [^] [4] [+] [2] [X,T,θ] [)] [^] [(] [1] [÷] [3] [)] [ENTER]	Enter $f(x)$ as Y1. `(X^4+2X)^(1/3)` Note that the [X,T,θ] key displays the letter X. (The letter T is displayed if parametric mode is being used, and θ is displayed if in polar graphing mode.)
[Y=] [CLEAR]	Clear the function stored as Y2.
[4] [^] [(] [2] [X,T,θ] [-] [90] [)] [2nd] [QUIT]	Enter $g(x)$ as Y2. `4^(2X-90)`
[45.2] [STO▸] [X,T,θ] [ENTER]	Store 45.2 as X.
[2nd] [Y-VARS] [1] <Function> [1] <Y1> [+] [3] [2nd] [√] [2nd] [Y-VARS] [1] <Function> [2] <Y2> [ENTER]	Form $f(x) + 3\sqrt{g(x)}$ and evaluate. The result is 164.97 to two decimal places.

EXAMPLE 3: Evaluate the expression $\frac{3x^2 + 5x - 8}{x^2 - 1}$ to three decimal places for $x = .900$, .990, .999, 1.001, 1.010, and 1.100 by: (A) evaluating the expression for each value of x individually; and (B) using a list.

Solution (A): First store the expression in the Y= list. Then store the value for x. Recall the expression and evaluate it.

Keystrokes	Explanation
[Y=] [CLEAR]	Clear the old expression for Y1 by pressing [CLEAR].
[(] [3] [X,T,θ] [^] [2] [+] [5] [X,T,θ] [–] [8] [)] [÷] [(] [X,T,θ] [^] [2] [–] [1] [)] [2nd] [QUIT]	Store the expression as Y1 and return to the home screen.
[.9] [STO▸] [X,T,θ] [ENTER]	Store the value of x.
[2nd] [Y-VARS] [1] <Function> [1] <Y1> [ENTER]	Recall the expression and evaluate.

Store the next value for x and reevaluate the expression. Repeat for the other values of x. The corresponding results are: 5.632, 5.513, 5.501, 5.499, 5.488, and 5.381.

Solution (B): Store the values of x in a list and evaluate the expression for each value of x simultaneously.

Keystrokes	Explanation
Y= CLEAR	Clear the old expression for Y1 by pressing CLEAR.
(3 X,T,θ ^ 2 + 5 X,T,θ − 8) ÷ (X,T,θ ^ 2 − 1) 2nd QUIT	Store the expression as Y1 and return to the home screen.
2nd { .9 , .99 , .999 , 1.001 , 1.01 , 1.1 2nd } STO▶ 2nd L1 ENTER	Store the values of x in the list L1.
2nd Y-VARS 1 <Function> 1 <Y1> (2nd L1) ENTER	Simultaneously calculate the value of the expression stored as Y1 for the values of x in list L1. The results are: 5.632, 5.513, 5.501, 5.499, 5.488, and 5.381. To view the results, use the ◀ and ▶ keys.

EXAMPLE 4: Evaluate the expression $\frac{3x^2 + 5x - 8}{x^2 - 1}$ to three decimal places for values of x at each tenth from 0 to 1 using a table.

Solution: First store the expression in the Y= list. Set the table parameters to begin at $x = 0$ and to have an increment of .1. Get the table.

Keystrokes	Explanation
Y= CLEAR	Clear the old expression for Y1 by pressing CLEAR.
(3 X,T,θ ^ 2 + 5 X,T,θ − 8) ÷ (X,T,θ ^ 2 − 1) 2nd QUIT	Store the expression as Y1 and return to the home screen.
2nd TblSet 0 ENTER .1 ENTER	Set the table to begin evaluating the expression at x = 0 with a step size of .1
ENTER ▼ ENTER	Set the calculator to automatically display values of x and Y1.
2nd TABLE ▼ ... ▼	Get the table. Arrow down to see the rest of the table. The highlighted value will appear at the bottom of the table.

X	Y1	
0	8	
.1	7.5455	
.2	7.1667	
.3	6.8462	
.4	6.5714	
.5	6.3333	
.6	6.125	
X=0		

A-5 Evaluating Inequalities

EXAMPLE 1: Given the inequality $x^2 - 3t < 5x + 9t^2$. Test this inequality for $x = 1.89$ and $t = -4.52$.

Solution:

Keystrokes	
[1.89] [STO▸] [X,T,θ] [ENTER]	Store the value of the variables.
[(-)] [4.52] [STO▸] [ALPHA]	
[T] [ENTER]	Enter the expression. X^2-3T<5X+9T^2
[X,T,θ] [^] [2] [-] [3] [ALPHA]	
[T] [2nd] [TEST] [5] << > [5]	
[X,T,θ] [+] [9] [ALPHA]	The result is 1. This means the inequality is true for these values of the variables. If the result was a 0 then the inequality would not have been true.
[T] [^] [2] [ENTER]	

A-6 Graphing

EXAMPLE 1: Graph $y = 3x^2-6$ and $y=3x^2+1$ on the same graphing screen.

Solution:

Keystrokes	
[Y=] [CLEAR]	Get the function list and clear the existing function for Y1. Enter the first function as Y1. Now clear the second function location and enter the second function as Y2.
[3] [X,T,θ] [^] [2] [-] [6] [ENTER]	
[CLEAR]	
[3] [X,T,θ] [^] [2] [+] [1] [ENTER]	Clear or deselect all others by placing the cursor over the = sign using the left arrow and pressing enter. Now this function will not graph. Note that the = sign is no longer highlighted.
[2nd] [DRAW] [1] <ClrDraw>	Clear drawings.
[ENTER]	
[2nd] [STAT PLOT] [1] [▶] [ENTER] etc.	Turn off all statistical plots.
[ZOOM] [6] <ZStandard>	The sixth option on the ZOOM menu will automatically set the graphing screen to have $-10 < x < 10$ and $-10 < y < 10$. Also, the functions will be automatically graphed.

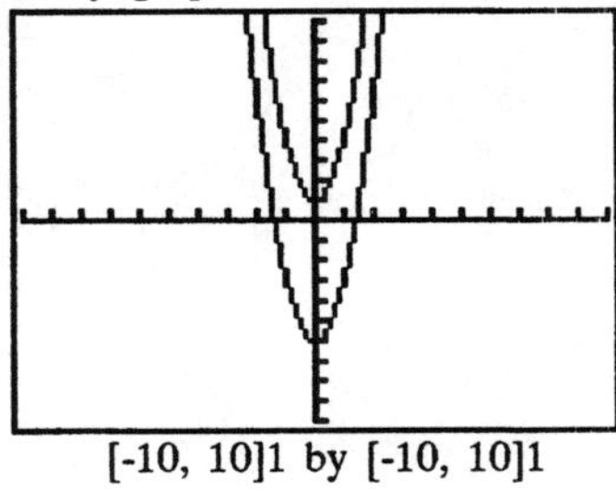

[-10, 10]1 by [-10, 10]1

| ZOOM | 6 | automatically sets the graphing screen limits to $-10 < x < 10$ and $-10 < y < 10$ with scale marks every unit for both the x and y axes. The graph screen limits will be denoted in this manual by the notation [-10, 10]1 by [-10, 10]1. The graph screen limits can be set to other values, if desired, by pressing | WINDOW | and entering the desired values, then pressing | GRAPH |.

| 6 | **<ZStandard>** means to choose the sixth option from the ZOOM menu.

The function will not graph if you move the cursor over the = sign and press | ENTER |. You notice the highlighting disappears from the = sign. You can still evaluate this expression for stored values of the variables. To select the function, place the cursor on the = sign and press | ENTER | again.

EXAMPLE 2: Graph $f(x) = \begin{cases} .5x + 1 & x \le 2 \\ x^2 + 2 & x > 2 \end{cases}$.

Solution: The calculator will only graph functions for values of x for which the function is defined. So we write this piecewise-defined function in a form where the pieces are defined only for the values of x that we want. Enter the function as two separate functions with (.5 X + 1) ÷ (X ≤ 2) as Y1 and (X ^ 2 + 2) ÷ (X > 2) as Y2. The first function is defined when $x \le 2$ and the second function is defined when $x > 2$. (See Appendix Section A-5 on how to get the inequality symbols from the | TEST | menu, and see Appendix Section A-6 Example 1 on how to store functions.)

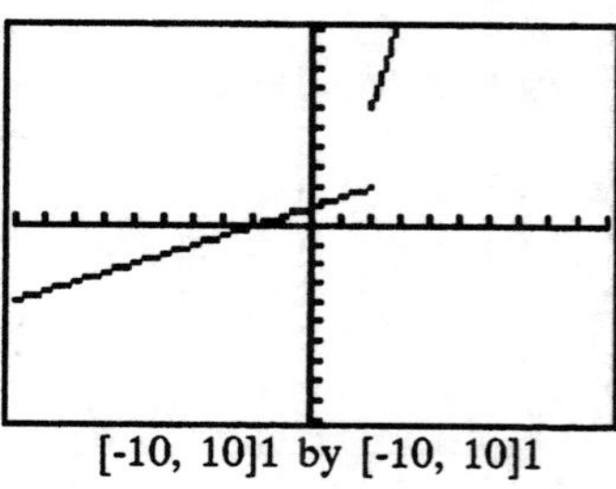

[-10, 10]1 by [-10, 10]1

A-7 TRACE, ZOOM, WINDOW, ROOT, and INTERSECT

EXAMPLE 1: Find the x intercept of $y = 3x^2 - 6$.

Solution:

First graph $y = 3x^2$-6 on the standard graphing screen (see Appendix Section A-6 Example 1 above).

| 2nd | CALC | 2 | <root>

After you get the root finder, you will see

Lower Bound?

on the calculator screen.

Use right and left arrow keys and move the cursor along the graph until the cursor is to the left of the root and press | ENTER |.

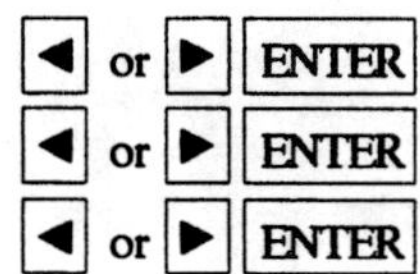

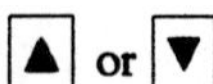

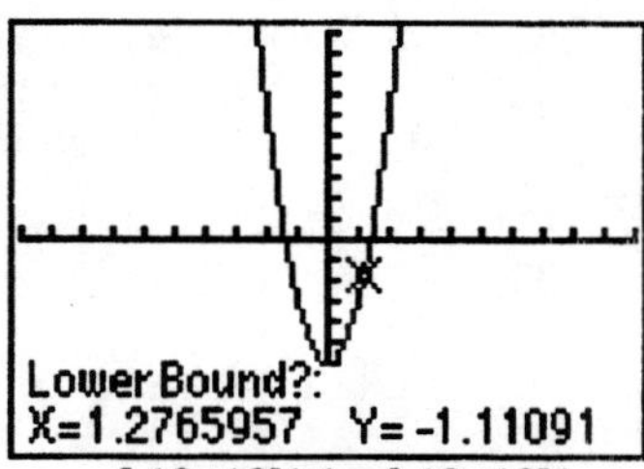

[-10, 10]1 by [-10, 10]1

Now move the cursor to the right of the room and press ENTER. Now move the cursor near the root and press ENTER. The calculator will display the x and y value of the root. Notice that the y value is very close to zero.

If we had two or more functions graphed, we could use the up and down arrow keys to move between the functions.

There are five ways to get a closer look at the intercept and one way to use the calculator to find the root:

1. Change the WINDOW dimensions.
2. Set the Zoom Factors and zoom in.
3. Use the Zoom Box feature of the calculator.
4. Use the Root Finder feature of the calculator.
5. Use the Intersect feature of the calculator.

EXAMPLE 2: Find the x intercept for $x>0$ of $y = 3x^2-6$. Use three decimal place accuracy.

Solution: Store the function in the calculator. Graph using [-10, 10]1 by [-10, 10]1.

<u>METHOD 1 Set the WINDOW dimensions.</u>

Tracing the function graphed above we see the x intercept is near 1.5. So we shall set the graph window close to that value.

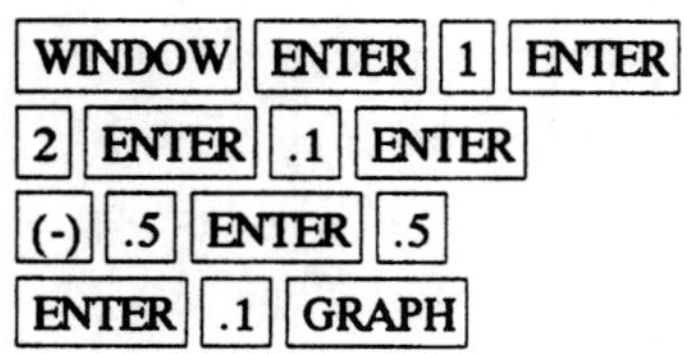

Set the WINDOW to [1, 2].1 by [-.5, .5].1. Notice we enter the minimum, maximum and scale values for the x axis, and then do the same for the y axis.

Use trace again to get a better approximation. We see that the x intercept is between 1.41 and 1.42. Set the WINDOW to [1.41, 1.42].01 by [-.01, .01].001. We now find the x intercept to be 1.414 accurate to three decimal places.

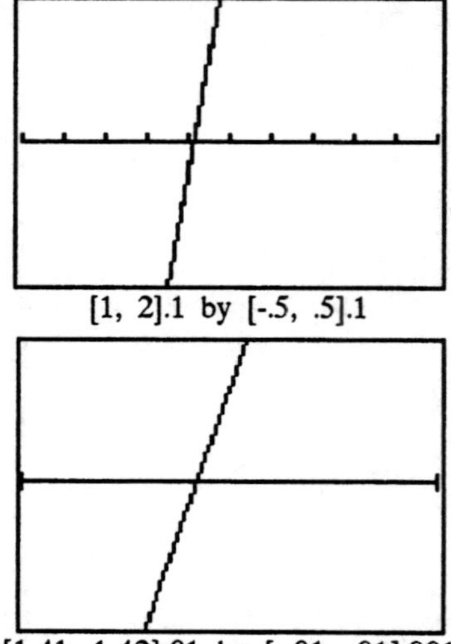

[1, 2].1 by [-.5, .5].1

[1.41, 1.42].01 by [-.01, .01].001

METHOD 2 Zoom In

ZOOM 6 <ZStandard>

ZOOM ► 4 <Set Factors>

5 ENTER 5 2nd QUIT

Get the [-10, 10]1 by [-10, 10]1 display as shown in Section A-6 Example 1. Choose the MEMORY from the ZOOM menu. Choose 4 <Set Factors>.

Set the Zoom Factors to 5 and 5. After zooming in the new Xmin will be $\frac{1}{5}$ the previous Xmin.

ZOOM 2 <Zoom In>.

◄ or ► or ▲ or ▼

ENTER

Use the arrow keys repeatedly to get the cursor close to the x intercept, say at $x=1.4893617$ $y=.64516129$. The new display will be centered where you placed the cursor.

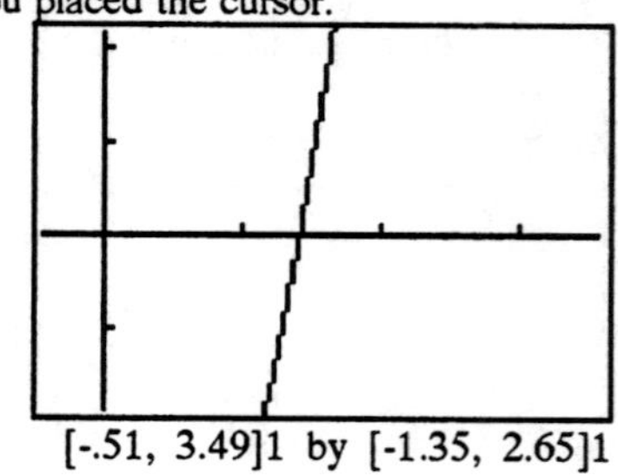

[-.51, 3.49]1 by [-1.35, 2.65]1

Now use TRACE again to get a new approximation to the x intercept.
Repeat zooming in as many times as necessary to get three decimal place accuracy.
The x intercept is 1.414.

METHOD 3 Zoom Box

Keystrokes	Explanation
ZOOM 6 <ZStandard>	Get the [-10, 10]1 by [-10, 10]1 display as shown in Section A-6 Example 1.
ZOOM 1 <ZBox>	Get the Zoom Box option.
◀ or ▶ or ▲ or ▼ ENTER	Use the arrow keys repeatedly to get the cursor close to but a little above and to the left of the x intercept, say at X=1.0638298 Y=.64516129. Press ENTER to place the upper corner of the Zoom Box at this point.
◀ or ▶ or ▲ or ▼ ENTER	Now move the cursor down and to the right of the x intercept, say at X = 1.7021277 Y =-.3225806. Press ENTER again. The graph will be drawn again with the box as the outside of the new display.

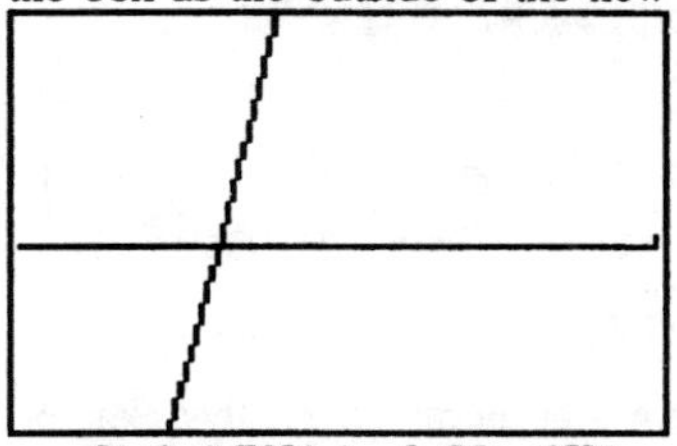

[1.6, 1.70]1 by [-.32, .65]1

Trace to get an estimate of the x intercept. Use Zoom Box a second time to get an even better approximation of the x intercept. The x intercept is 1.414.

METHOD 4 Root Finder

Keystrokes	Explanation
2nd CALC 2 <root>	
◀ or ▶ ENTER	To place the cursor at a point on the graph to the left of the x intercept.
◀ or ▶ ENTER	To place the cursor at a point on the graph to the right of the x intercept.
ENTER	For the guess.

The value of the root will appear at the bottom of the screen. The root is (1.41, 0) rounded to two decimal places. Repeat for the other root.

METHOD 5 Intersect

Keystrokes	Explanation
Y= ▼ 0 2nd QUIT	To enter $y = 0$ as the second function.
2nd CALC 5 <intersect>	
◀ or ▶ ENTER	To place the cursor at a point on the graph of the first curve near one of the x intercepts.
◀ or ▶ ENTER	To place the cursor at a point on the graph of the second curve near the same intercept.
◀ or ▶ ENTER	For the guess.

The coordinates of the point of intersection will appear at the bottom of the screen. The coordinates of the rightmost x intercept is (1.41, 0) rounded to two decimal places. Repeat for the other x intercept.

A-8 Choosing the WINDOW Variables Values

There is no one right way to determine the window variables that should be used for the graph of a function. Three methods to try are:

1. Graph using the standard screen and then zoom in or zoom out, if necessary.
2. Analyze the features of the function and then choose the viewing window dimensions.
3. Evaluate the function at several values and then choose the viewing window dimensions.

EXAMPLE 1 Graph $y = .03x^3 + .02x^2 - 5x + 10$.

Solution:

METHOD 1 Graph using the standard screen and then zoom out

Enter the function as .03 X ^ 3 + .02 X ^ 2 - 5 X + 10. Press [ZOOM] [6] to get the [-10, 10]1 by [-10, 10]1 display. (See Section A-6 of this manual.) We have too small a display to see the behavior of the graph clearly. Press [ZOOM] [▶] [4] and set the Zoom Factors to XFact=4 and YFact=4. Press [ZOOM] [3], move the cursor so it is where you want the center of the new screen to be, (we will leave it at the origin for this example) and press [ENTER]. The result is shown below. Note the double lines as the axes. This is because the Xscl and Yscl values do not change when zooming in. Since the scale values for x and y are small compared to the distance between Xmin and Xmax, and Ymin and Ymax, the scale marks appear to run together. You can fix this by choosing a more appropriate scale (say at 5 for Xscl and Yscl for this example). [Note if Xscl or Yscl are set to 0, no scale marks will be drawn.]

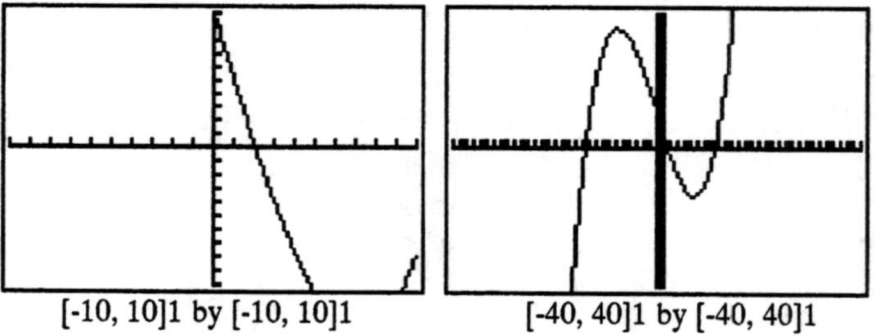

[-10, 10]1 by [-10, 10]1   [-40, 40]1 by [-40, 40]1

METHOD 2 Analyze the function and then choose the window variable values

We will analyze the term with highest power on x since this is the dominant term of the function.

We see that the leading coefficient is .03. This means that the vertical change in the graph will be approximately .03 unit for every time x^3 increases 1 unit. An equation that represents this is $.03x^3 = y$ or $x^3 = \frac{1}{.03}y$ or $x^3 \approx 30y$. So if y increases by 1 unit, x will increase by about $\sqrt[3]{30}$ or about 3 units. This indicates that we want the x axis three times as wide as the y axis. Hence, a good first choice for the horizontal axis is -30 to 30, and a good first choice for the vertical axis is -10 to 10. Set Xmin = -30, Xmax = 30 and Xscl = 3. We choose 3 as the Xscl because this will put 20 tick marks on the axis.

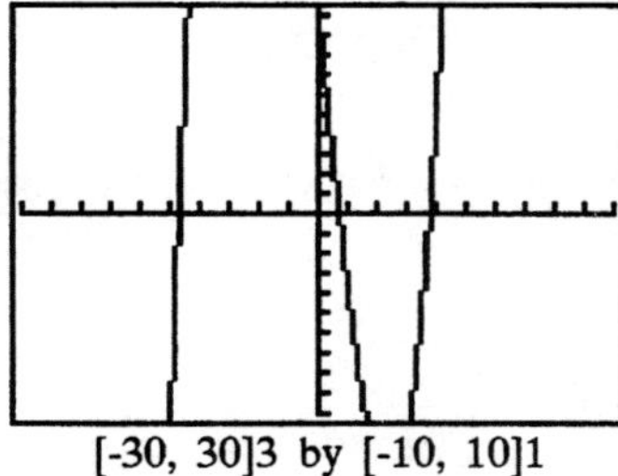

[-30, 30]3 by [-10, 10]1

The values for the vertical variables needs some adjusting since the turning points of the graph are not shown on the screen. Window variable values of [-30, 30]1 by [-40, 40]5 would show the turning points of the graph.

METHOD 3 Evaluate the function at several values and then choose the viewing window dimensions

Evaluate the function by storing a value for x and evaluating the function (see Section A-4). Repeat for each value of x. The choice of x values is arbitrary.

x	$f(x)$
-25	-321.25
-10	32
0	10
10	-8
25	366.25

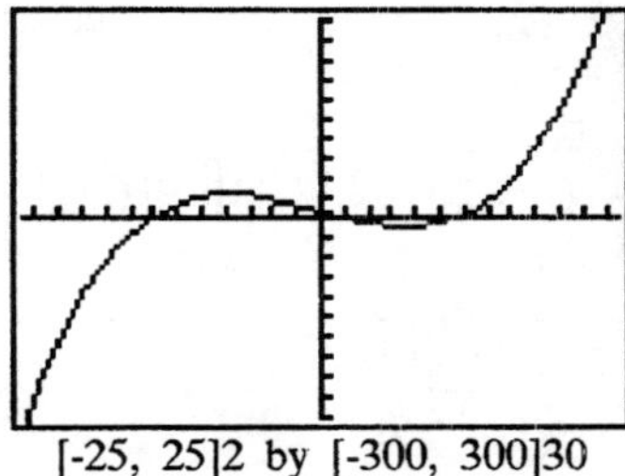

[-25, 25]2 by [-300, 300]30

Choose the Xscl and Yscl so there are about 20 marks on each axis (see METHOD 2 above). So a good first choice for the WINDOW could be [-25, 25]2 by [-300, 300]30.

A-9 Equations in One Variable

An equation in one variable can be solved graphically in two ways:

1. Algebraically move all terms to one side of the equation. Create a function $f(x)$=(the expression). Graph the function. Zoom and trace, or use other calculator features, to find the x intercepts. (See Appendix Section A-7 of this manual.) The set of x intercepts is the solution to the equation.

2. Create two functions: $f(x)$ = (the left side of the equation) and $g(x)$ = (the right side of the equation). Graph both functions. Zoom and trace, or use other calculator features, to find the x value of the points of intersection of the two graphs. (See Appendix Section A-7 of this manual.) The set of x values is the solution to the equation.

EXAMPLE: Solve $\sqrt[3]{x+2} = 4x^2 - 3\sqrt{x+4}$. Use two decimal places.

Solution:

METHOD 1 Write the equation as $\sqrt[3]{x+2} - (4x^2 - 3\sqrt{x+4}) = 0$. Create the function $f(x) = \sqrt[3]{x+2} - (4x^2 - 3\sqrt{x+4})$. Enter this function into the calculator and graph.

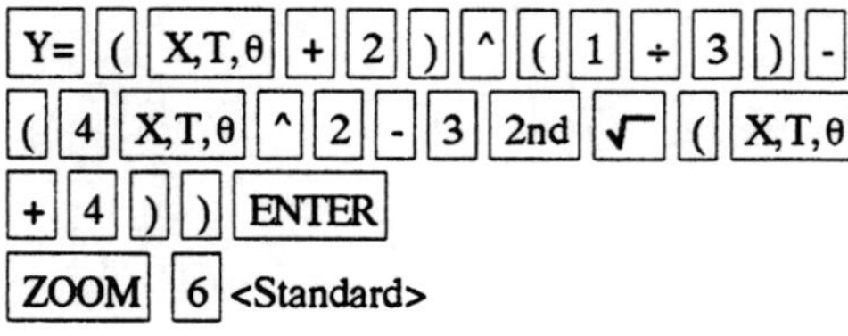

<Standard>

Enter the function and graph.

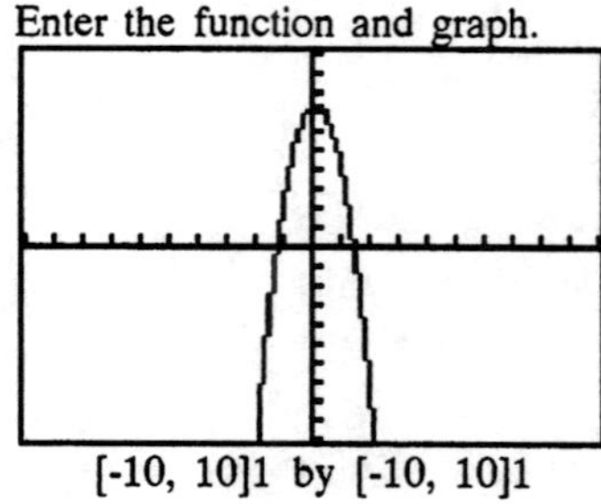

[-10, 10]1 by [-10, 10]1

We see the function crosses the x axis in two places. Trace and zoom, or use other calculator features, to find these x values. The solution to this equation is $x = -1.22$ and $x = 1.46$.

METHOD 2 Create the two functions $f(x) = \sqrt[3]{x+2}$ and $g(x) = 4x^2 - 3\sqrt{x+4}$. Graph these and find the x values of the intersection points.

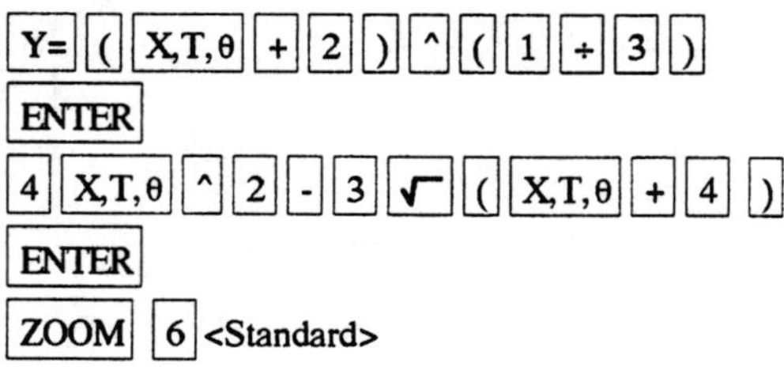

<Standard>

Enter the functions and graph.

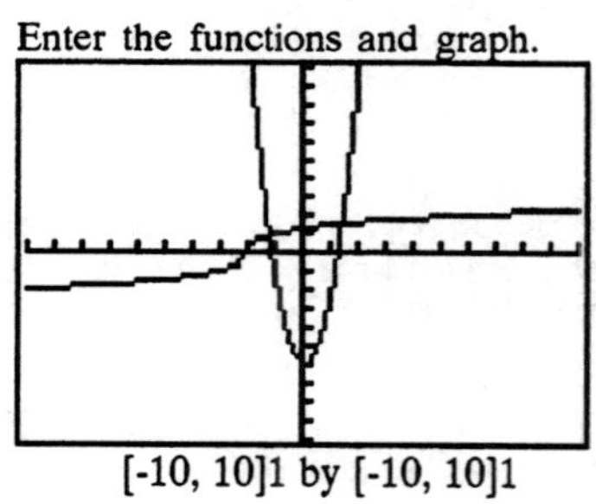

[-10, 10]1 by [-10, 10]1

We see the functions intersect in two places. Trace and zoom to find the coordinates of these points. The intersection points are approximately at (-1.22, .92) and (1.46, 1.51). So the solution to this equation is $x = -1.22$ and $x = 1.46$.

A-10 Inequalities in One Variable

An inequality in one variable can be solved graphically in two ways:

1. Algebraically move all terms to one side of the inequality and change the sign to an equal sign. Create a function $f(x)$ = (the expression). Graph the function. Trace and zoom to find the x intercepts. The set of x intercepts are the cutoff values for the solution. Now determine the intervals over which the inequality is true and write the solution. Pick a value either greater or less than the x intercept. Test the inequality using the method shown in Appendix Section A-5.

2. Create two functions: $f(x)$ = (the left side of the inequality) and $g(x)$ = (the right side of the inequality). Graph both functions. Trace and zoom to find the x value of the points of intersection of the two graphs. The set of x coordinates are the cutoff values for the solution. Now determine the intervals over which the inequality is true and write the solution. Pick a value either greater or less than the x intercept. Test the inequality using the method shown in Appendix Section A-5.

EXAMPLE: Solve $\sqrt[3]{x+2} < 4x^2 - 3\sqrt{x+4}$. Use two decimal places

Solution:

<u>METHOD 1</u> Write the inequality as: $\sqrt[3]{x+2} - (4x^2 - 3\sqrt{x+4}) < 0$. Change the sign to an equal sign: $\sqrt[3]{x+2} - (4x^2 - 3\sqrt{x+4}) = 0$. Create the function $f(x) = \sqrt[3]{x+2} - (4x^2 - 3\sqrt{x+4})$, graph, and find the x intercepts. This was done above in Appendix Section A-9 METHOD 1 of this manual. The cutoff values are $x = -1.22$ and $x = 1.46$. Observing the graph we see it is below the x axis for $x < -1.22$ and $x > 1.46$. Thus for these values of x we have $f(x) < 0$ and hence $\sqrt[3]{x+2} < 4x^2 - 3\sqrt{x+4}$. Hence the solution to this inequality can be written as $(-\infty, -1.22) \cup (1.46, \infty)$.

<u>METHOD 2</u> Create the two functions $f(x) = \sqrt[3]{x+2}$ and $g(x) = 4x^2 - 3\sqrt{x+4}$. Graph these and find the x values of the intersection points. This was done above in Appendix Section A-9 METHOD 2 of this manual. The intersection points are at (-1.22, .92) and (1.46, 1.52). Inspecting the graph we see that $f(x) < g(x)$ for $x < -1.22$ and $x > 1.46$. Hence the solution to this inequality can be written as $(-\infty, -1.22) \cup (1.46, \infty)$.

A-11 Permutations, Combinations, and Random Numbers

EXAMPLE 1: Find $P_{19,13}$ the number of permutations of 19 distinct objects taken 13 at a time without repetition.

Solution:

Keystrokes	Explanation
[19] [MATH] [►] [►] [►] [2] <nPr> [13] [ENTER]	Enter the numbers and get the built-in permutation function from the MATH menu. The result is displayed in scientific notation as 1.689515283E14.

EXAMPLE 2: Find $C_{19,13}$ the number of combinations of 19 distinct objects taken 13 at a time without repetition.

Solution:

Keystrokes	Explanation
[19] [MATH] [►] [►] [►] [3] <nCr> [13] [ENTER]	Enter the numbers and get the built-in function from the MATH menu. The result is 27132.

EXAMPLE 3: Generate ten unique random numbers at least 0 and less than 50.

Solution:

[5] [STO▸] [MATH] [▸] [▸] [▸]
[1] <rand> [ENTER]

We need to store a "seed" number in the storage location RAND. This tells the random number generator where to start.

[MATH] [▸] [2] <iPart> [(]
[50] [x] [MATH] [▸] [▸] [▸]
[1] <rand> [)] [ENTER] [ENTER]

RAND generates a number greater than 0 and less than 1. If we multiply by 50 and take the integer part, the result is a number greater than or equal to 0 and less than 50.

[ENTER] [ENTER] [ENTER] [ENTER]
[ENTER] [ENTER] [ENTER] [ENTER]

The 10 random numbers are:

36 13 6 17 21 ~~21~~ 20 5 31 16 48

Note that repetitions can occur but 21 cannot be used twice.

A-12 Matrices

Given the matrices $A = \begin{bmatrix} 1 & -4 & 6 \\ 2 & -1 & 0 \\ 5 & -3 & 6 \end{bmatrix}$ and $B = \begin{bmatrix} 4 & 15 \\ -2 & 6 \\ 2 & 5 \end{bmatrix}$.

EXAMPLE 1: Find -2.5AB.

Solution:

[MATRX] [▶] [▶] [1] <[A]> [3]
[ENTER] [3] [ENTER]

Choose matrix A and enter the dimensions.

[1] [ENTER] [(-)] [4] [ENTER]
[6] [ENTER] [2] [ENTER]
[(-)] [1] [ENTER] [0] [ENTER]
[5] [ENTER] [(-)] [3]
[ENTER] [6] [ENTER]

Enter the matrix elements.

[MATRX] [▶] [▶] [2] <[B]>
[3] [ENTER] [2] [ENTER]

Choose matrix B and enter the dimensions.

[4] [ENTER] [15] [ENTER]
[(-)] [2] [ENTER] [6] [ENTER]
[2] [ENTER] [5]

Enter the matrix elements.

[2nd] [QUIT]

Return to the Home Screen. Note that [A] is the matrix A. This is different than A which represents the variable A.

[(-)] [2.5] [MATRX] [1] <[A]> [x] [MATRX] [2] <[B]> [ENTER]

Enter the matrix operations.

The result is $\begin{bmatrix} -60 & -52.5 \\ -25 & -60 \\ -95 & -217.5 \end{bmatrix}$.

EXAMPLE 2: Find A^{-1} using matrix A from Example 1.

Solution:

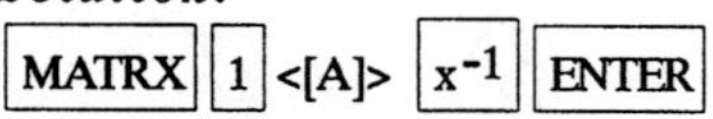

The operations are done from the Home Screen. The result rounded to three decimal places is:

$$A = \begin{bmatrix} -.167 & .167 & .167 \\ -.333 & -.667 & .333 \\ -.028 & -.472 & .194 \end{bmatrix}$$

[STO▸] [MATRX] [2] <[B]> [ENTER]

To inspect the result, store it as another matrix, say [B], recall it and use the arrow keys to see the elements.

EXAMPLE 3: Solve the system of equations using the inverse of the coefficient matrix. Round answers to two decimal places.

$$\begin{aligned} x - 4y + 6z &= 2.5 \\ 2x - y \quad\quad &= 3.2 \\ 5x - 3y + 6z &= -3.4 \end{aligned}$$

Solution: Note that the coefficient matrix is matrix A from Section A-12 Example 1 of this manual. The method of finding the inverse of a matrix was shown above in Section A-12 Example 2. Enter the constants as a 3×1 matrix B and find $A^{-1}B$.

[MATRX] [▶] [▶] [2] <[B]> [3] [ENTER] [1] [ENTER]

Enter the matrix dimensions.

[2.5] [ENTER] [3.2] [ENTER] [(-)] [3.4]

Enter the matrix elements.

[2nd] [QUIT]

Return to the Home Screen.

[MATRX] [1] <[A]> [x^{-1}] [MATRX] [2] <[B]> [ENTER]

Perform the matrix operations.

The result is $\begin{bmatrix} -.45 \\ -4.10 \\ -2.24 \end{bmatrix}$. The solution to the system of equations to two decimal places is $x = -.45$, $y = -4.10$, $z = -2.24$.

EXAMPLE 4: Find the solution to the system of Example 3 above using Gauss-Jordan reduction. In other words, find the reduced form of the matrix $\begin{bmatrix} 1 & -4 & 6 & 2.5 \\ 2 & -1 & 0 & 3.2 \\ 5 & -3 & 6 & -3.4 \end{bmatrix}$.

Solution:

Keystrokes	Explanation
[MATRX] [▶] [▶] [1] <[A]>	Name the matrix A. Enter the matrix dimensions.
[3] [ENTER] [4] [ENTER]	
[1] [ENTER] [(-)] [4] [ENTER]	Enter the matrix elements row by row.
[6] [ENTER] [2.5] [ENTER]	
[2] [ENTER] [(-)] [1] [ENTER]	
[0] [ENTER] [3.2] [ENTER]	
[5] [ENTER] [(-)] [3] [ENTER]	
[6] [ENTER] [(-)] [3.4]	
[2nd] [QUIT]	Return to the home screen.

After a calculation, some numbers may look like decimals. However upon storing, recalling and examining the matrix using the arrow keys you will find they are decimals that are very close to zero. This situation can be eliminated by setting the number of decimal places to a number, say 3 (see Section A-3 of this manual).

Keystrokes	Explanation
[MATRX] [1] <[A]> [ENTER]	Display the matrix.
[MATRX] [▶] [ALPHA] [A] <*row+(> [(-)] [2] [,] [MATRIX] [1] <[A]> [,] [1] [,] [2] [)] [ENTER]	We wish to multiply by -2, matrix A, row 1 and add it to row 2. The result will be put in row 2. `*row+(-2,[A],1,2)` The result is $\begin{bmatrix} 1 & -4 & 6 & 2.5 \\ 0 & 7 & -12 & -1.8 \\ 5 & -3 & 6 & -3.4 \end{bmatrix}$.
[STO▶] [MATRIX] [1] <[A]> [ENTER]	Store the result as matrix A so you can perform the next operation.
[MATRX] [▶] [ALPHA] [A] <*row+(> [(-)] [5] [,] [MATRIX] [1] <[A]> [,] [1] [,] [3] [)] [ENTER]	Now multiply by -5, matrix A, row 1 and add it to row 3. `*row+(-5,[A],1,3)` The result is $\begin{bmatrix} 1 & -4 & 6 & 2.5 \\ 0 & 7 & -12 & -1.8 \\ 0 & 17 & -24 & -15.9 \end{bmatrix}$. Store the result as matrix A.
[STO▶] [MATRIX] [1] <[A]> [ENTER]	

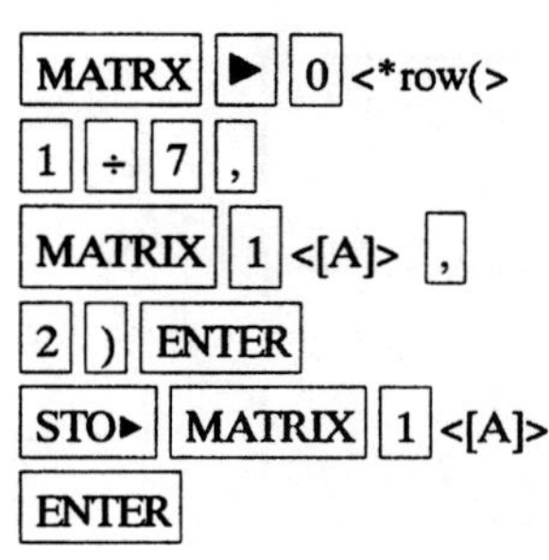

Multiply by 1/7, matrix A, row 2.

```
*row(1÷7,[A],2)
```

The result is $\begin{bmatrix} 1 & -4 & 6 & 2.5 \\ 0 & 1 & -1.71 & -.26 \\ 0 & 17 & -24 & -15.9 \end{bmatrix}$ rounded to two decimal places.
Store the result as matrix A.

Continue in this manner until you get $\begin{bmatrix} 1 & 0 & 0 & -.45 \\ 0 & 1 & 0 & -4.10 \\ 0 & 0 & 1 & -2.24 \end{bmatrix}$. The solution to the system of equations is $x = -.45$, $y = -4.10$, $z = -2.24$.

> Note: To swap rows of a matrix use [8] <rowSwap(> from the [MATRX] MATH menu. The command rowSwap([A],2,3) will swap rows 2 and 3 in matrix A.
>
> Note: To add one row to another use [9] <row+(> from the [MATRX] MATH menu. The command row+([A],2,3) will add row 2 to row 3 in matrix A.

A-13 Logarithmic, Exponential, and Hyperbolic Functions

EXAMPLE 1: Graph $f(x) = 10^{3.2x}$ for $-.5 < x < .5$.

Solution: Set the WINDOW to [-.5, .5].1 by [-10, 10]1. (See Appendix B Section A-7 Example 2 Method 1.)

[Y=] [10] [^] [(] [3.2] [X,T,θ] [)] [GRAPH]

Enter the functions and graph.

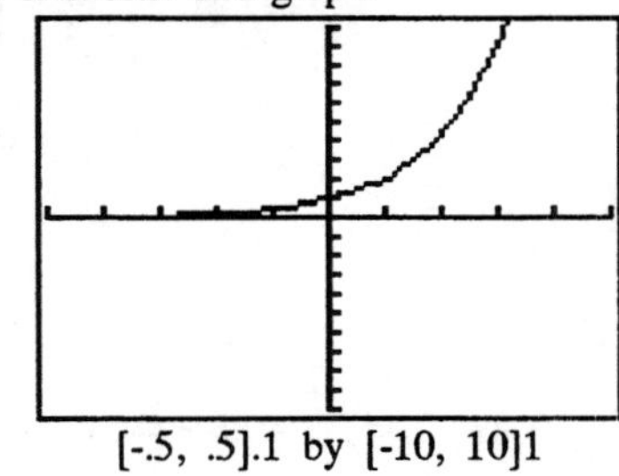

[-.5, .5].1 by [-10, 10]1

EXAMPLE 2: Graph $y = \dfrac{e^{x} + e^{-x}}{2}$.

Solution: This could be entered using [(] [2nd] [e^x] [X,T,θ] [+] [2nd] [e^x] [(-)] [X,T,θ] [)] [÷] [2]. However this is the hyperbolic cosine function. The following uses this built-in function. Recall that [e^x] represents the built-in exponential function above the [LN] key.

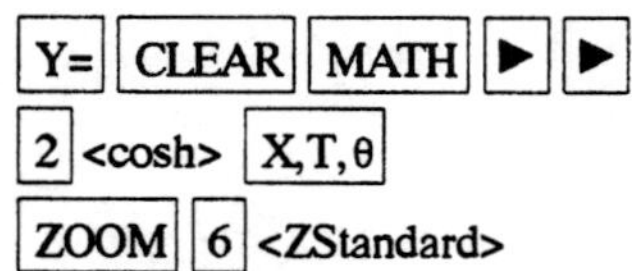

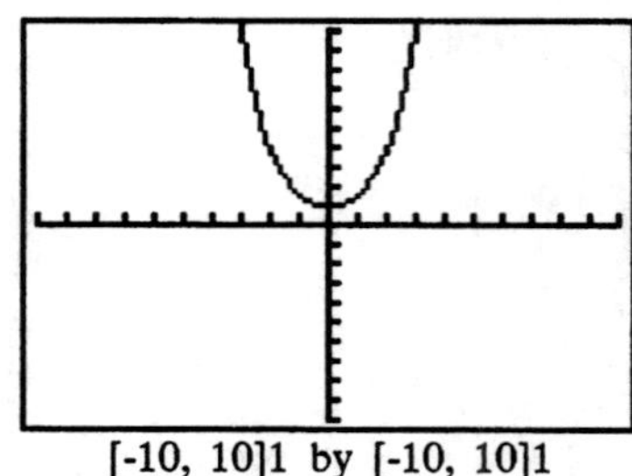

[-10, 10]1 by [-10, 10]1

EXAMPLE 3: Solve $\ln(x^2 + x) = 3x + 2$.

Solution:

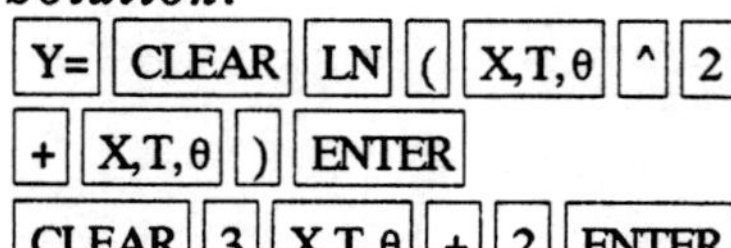

Enter the function $f(x) = \ln(x^2 + x)$ as Y1 in the function list.

[CLEAR] [3] [X,T,θ] [+] [2] [ENTER]

Enter the function $g(x) = 3x + 2$ as Y2 in the function list.

[ZOOM] [6] <ZStandard>

Graph them on the same graphing screen. Trace and zoom to find the x value of the point of intersection. This will be the solution to the equation.

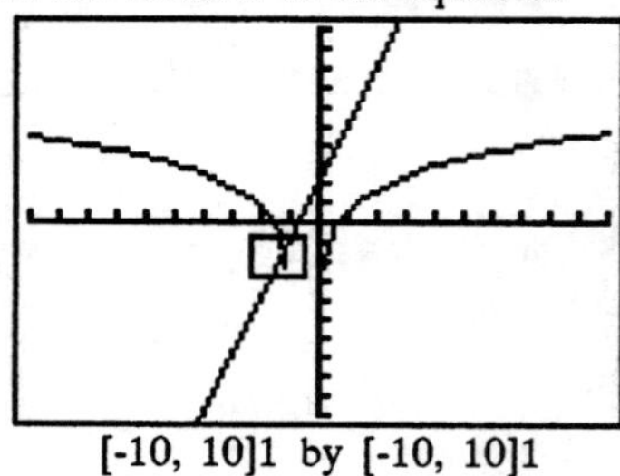

[-10, 10]1 by [-10, 10]1

The solution is -1.18, rounded to two decimal places.

A-14 Trigonometric Functions

EXAMPLE 1: Evaluate $f(x) = \cos x + \csc x$ at: (A) 53°45'18" and (B) 78.52 radians

Solution: (A)

[Y=] [CLEAR] [COS] [X,T,θ] [+] [1]
[÷] [SIN] [X,T,θ] [ENTER]

Store the function.

[MODE] [▼] [▼] [▶] [ENTER]

Change the calculator to degree angle measure.

[2nd] [QUIT]

[53] [+] [45] [÷] [60] [+] [18] [÷]
[3600] [STO▶] [X,T,θ]

Change the angle to decimal degrees while storing it as the variable X.

[2nd] [Y-VARS] [1] [1] <Y1>
[ENTER]

Evaluate the function. The result to two decimal places is 1.83.

Solution: (B)

[MODE] [▼] [▼] [ENTER]
[2nd] [QUIT]

Change the calculator to radian angle measure. Since the function was stored in Part (A) above, we need only to store the angle measure and evaluate the stored function.

[78.52] [STO▸] [X,T,θ] [ENTER]

Store 78.52 as X.

[2nd] [Y-VARS] [1] [1] <Y1>
[ENTER]

Evaluate the function. The result to two decimal places is 49.47.

EXAMPLE 2: Graph $f(x) = \cos x + \csc x$.

Solution: The function was entered in the calculator in Part (A) above.

[MODE] [▼] [▼] [▼] [▼] [▶]
[ENTER] [2nd] [QUIT]

Change the graph mode to Dot. This will eliminate the "vertical lines."

[ZOOM] [7] <ZTrig>

This option automatically sets the viewing window variables to [-6.28, 6.28]1.57 by [-3, 3].25.

[-6.28, 6.28]1.57 by [-3, 3].25

A-15 Numeric Differentiation, Integration, and Tangent Lines

EXAMPLE 1: (A) Find the equation of the tangent line to $f(x) = \sqrt{x^5 - \sqrt[3]{x}}$ at $x = 3.54$. Use two decimal place accuracy. (B) Sketch the function and its tangent line using [2, 5]1 by [0, 60]5.

Solution (A): The function and its derivative can be evaluated at this value of x.

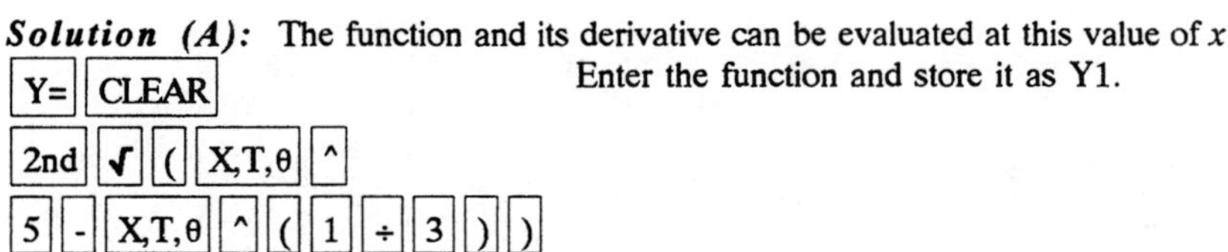

[Y=] [CLEAR]
[2nd] [√] [(] [X,T,θ] [^]
[5] [-] [X,T,θ] [^] [(] [1] [÷] [3] [)] [)]

Enter the function and store it as Y1.

Keystrokes	Explanation
[2nd] [QUIT]	Get the Home Screen and store the value of x.
[3.54] [STO▶] [X,T,θ] [ENTER]	
[2nd] [Y-VARS] [1] [1] <Y1>	Evaluate the function. $f(3.54) = 23.55$ to two decimal places.
[ENTER]	
[MATH] [8] <nDeriv(>	Get the numeric derivative built-in function. `nDeriv(Y1,x,3.54,.001)`
[2nd] [Y-VARS] [1] [1] <Y1>	This will evaluate the derivative with respect to x of
[,] [X,T,θ] [,] [3.54] [,] [.0001] [)]	the function stored in Y1 at $x = 3.54$ with .0001 tolerance. If the tolerance is not entered, the calculator will use .001.
[ENTER]	The result is 16.67.

The equation of the tangent line is $y - 23.55 = 16.67(x - 3.54)$.

Solution (B): Method 1 Enter the function as Y1 in the calculator and the equation of the tangent line as Y2 in the calculator and graph.

Method 2 Enter the function as Y1 in the calculator. Set the viewing rectangle to [2, 5]1 by [0, 60]5. Use the tangent drawing feature of the calculator by entering

[2nd] [QUIT] [2nd] [DRAW] [5] <Tangent(> [2nd] [1] <Function> [1] <Y1> [,] [3.54] [)] [ENTER]. The function and its tangent will be graphed as shown to the right.

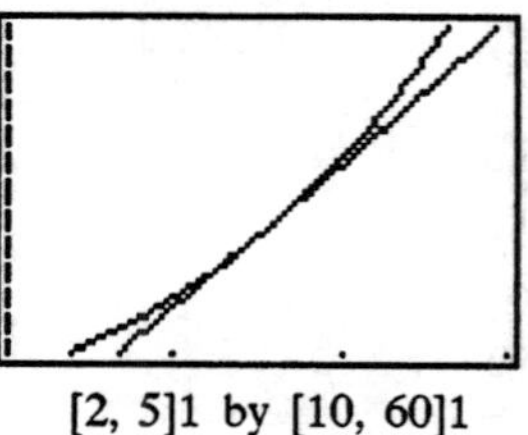

[2, 5]1 by [10, 60]1

EXAMPLE 2: Find $\int_{1.32}^{2.58} \sqrt{x^5 - \sqrt[3]{x}}\, dx$. Use two decimal place accuracy.

Solution:

Keystrokes	Explanation
[Y=] [CLEAR] [2nd] [√] [(] [X,T,θ] [^]	Enter the function and store it.
[5] [-] [X,T,θ] [^] [(] [1] [÷] [3] [)] [)]	
[2nd] [QUIT]	Get the numeric integration function built into the calculator function:
[MATH] [9] <fnInt(>	`fnInt(Y1,X,1.32,2.58).`
[2nd] [Y-VARS] [1] [1] <Y1>	This will evaluate the definite integral. The result is 6.95. The tolerance in calculating
[,] [X,T,θ] [,] [1.32] [,] [2.58] [)]	the definite integral is .00001.

A-16 Statistics

EXAMPLE 1: Given the following data find: (A) the mean; (B) the sample standard deviation ; (C) an histogram with first class beginning at 0 and width 4; (D) repeat Parts A and B after deleting the data value 89 from the list. How did eliminating this outlier effect the mean and standard deviation?

23	56	73	12	89	52	45	19	40	57	39
38	18	21	15	37	42	51	38	54	60	

Use one decimal place in answers.

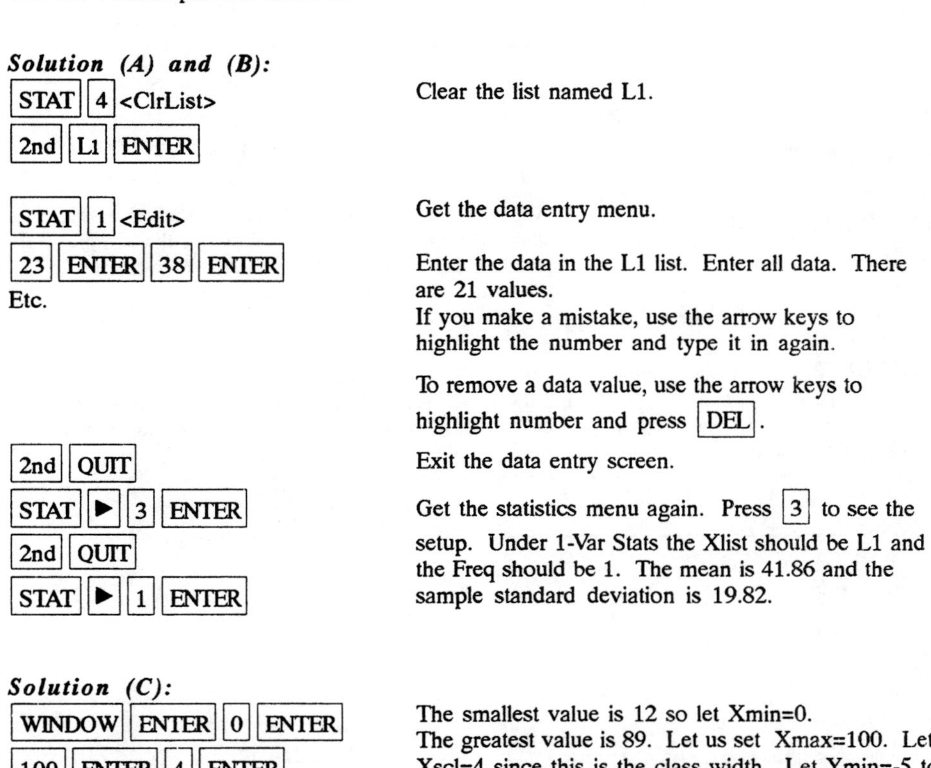

Solution (A) and (B):

Keystrokes	Explanation
[STAT] [4] <ClrList> [2nd] [L1] [ENTER]	Clear the list named L1.
[STAT] [1] <Edit>	Get the data entry menu.
[23] [ENTER] [38] [ENTER] Etc.	Enter the data in the L1 list. Enter all data. There are 21 values. If you make a mistake, use the arrow keys to highlight the number and type it in again. To remove a data value, use the arrow keys to highlight number and press [DEL].
[2nd] [QUIT]	Exit the data entry screen.
[STAT] [►] [3] [ENTER] [2nd] [QUIT] [STAT] [►] [1] [ENTER]	Get the statistics menu again. Press [3] to see the setup. Under 1-Var Stats the Xlist should be L1 and the Freq should be 1. The mean is 41.86 and the sample standard deviation is 19.82.

Solution (C):

Keystrokes	Explanation
[WINDOW] [ENTER] [0] [ENTER] [100] [ENTER] [4] [ENTER] [(-)] [5] [ENTER] [5] [ENTER] [1]	The smallest value is 12 so let Xmin=0. The greatest value is 89. Let us set Xmax=100. Let Xscl=4 since this is the class width. Let Ymin=-5 to get some room below the histogram. We do not know the frequencies yet so let Ymax=5 for a first try. Let Yscl=1.
[Y=] [CLEAR] ...	Clear all functions on the function list.
[2nd] [DRAW] [1] <ClrDraw> [ENTER]	Clear all drawings from the display.

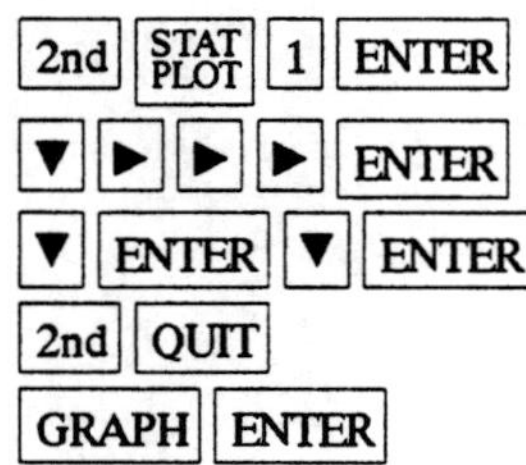

Set PLOT 1 to ON.
Select Type as Histogram.
Select Xlist as L1 and Freq as 1.
Graph the histogram.

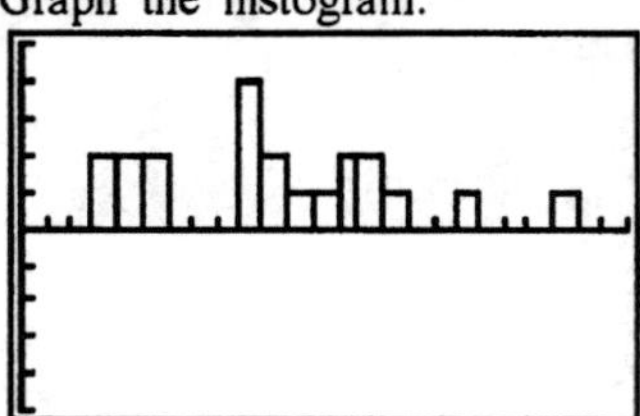

Solution (D):

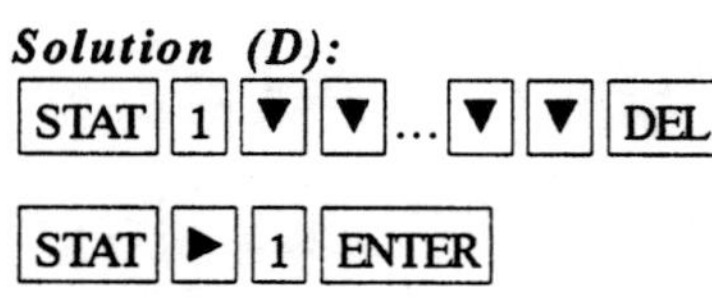

Get the data list and arrow down to the 89. Press DEL.

Get the statistics menu and the one variable sample statistics. The mean is now 39.5 and the sample standard deviation is 17.05. Notice how much influence an outlier has!

EXAMPLE 2: Given the following bivariate data: (A) Find the correlation coefficient. (B) Find the least squares line. (C) Graph the least squares line and the scatter plot on the same graph. (D) Plot the scatter diagram without the least squares line. (E) Sort the data on x and join the points using straight line segments.

x	5	8	10	29	9	24
y	3	4	8	20	3	19

Solution (A):

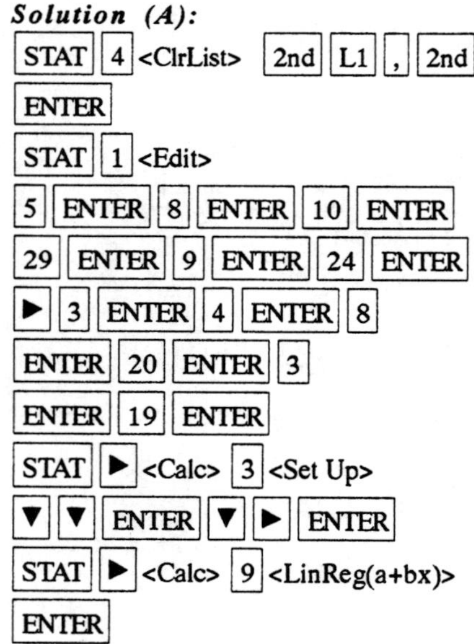

Clear list L1 and list L2.

Get the lists.

Enter the data in list L1.

Move to list L2. Enter the data in list L2. The data for y in L2 must correspond pairwise to the data for x in L1.

Set up the calculator to use L1 as the x variable and L2 as the y variable in the calculations.

This give you the linear correlation coefficient, the constant term and the coefficient on x term for the least squares line. $r = .98$.

Solution (B):

VARS 5 <Statistics> ► ►
7 <RegEQ>

The least squares line equation is displayed. It is $\hat{y}$ = -1.72 + .79*x* with the coefficients rounded to two decimal places.

Solution (C):

WINDOW ENTER 0 ENTER 30
ENTER 1 ENTER (-) 5 ENTER
30 ENTER 1

Set the graph WINDOW to [0, 30]1 by [-5, 30]1. Observe the data to determine the WINDOW values. Set the *y* value lower than needed so if you trace, the coordinates at the bottom of the screen do not cover up the graph.

Y= CLEAR ...

Clear all functions on the function list.

2nd DRAW 1 <ClrDraw> ENTER

Clear all drawings from the display.

Turn the other STAT PLOTS off.

Y= VARS 5 ► ►
7 <RegEQ>

The least squares line equation will be put into the function list so you can graph it.

2nd STAT PLOT 1 <Plot 1> ENTER
▼ ENTER ▼ ENTER ▼ ►
ENTER ▼ ENTER
GRAPH

Set up the calculator to get the scatter plot with the line.

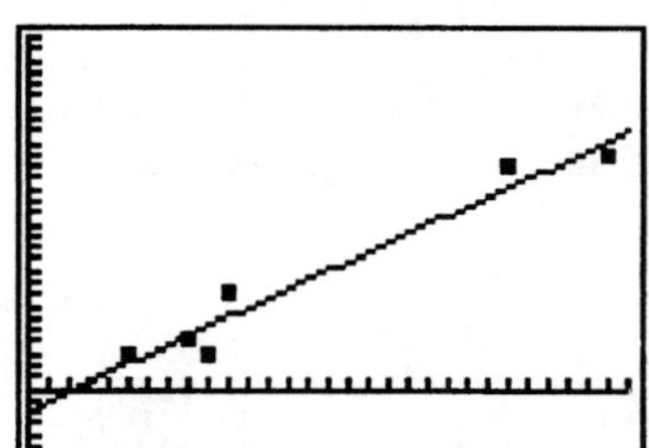

Solution (D):

Y= CLEAR or Y= ◄ ENTER

Clear the functions from the list or deselect them so they won't graph.

GRAPH

Plot the points on the graph.

Solution (E):

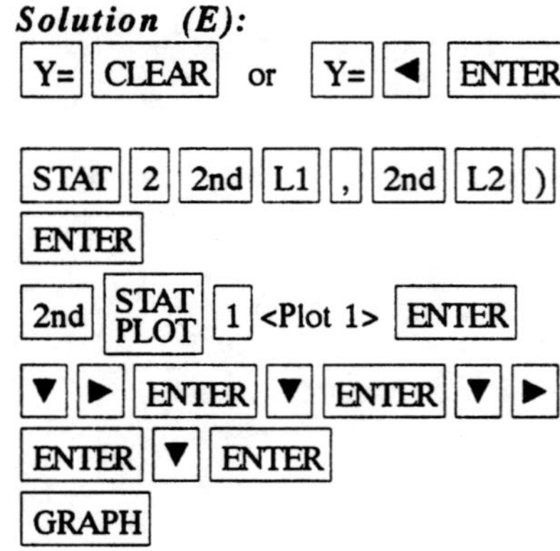

Clear the functions from the list or deselect them so they won't graph.

Sort the data on the x variable. The list L1 will be sorted in ascending order. The list L2 will sort along with L1 as pairs.
Set the calculator to draw line segments joining the points.
This will join the points together with straight line segments in the order they appear in the data list. This is why we sorted the data on x first.

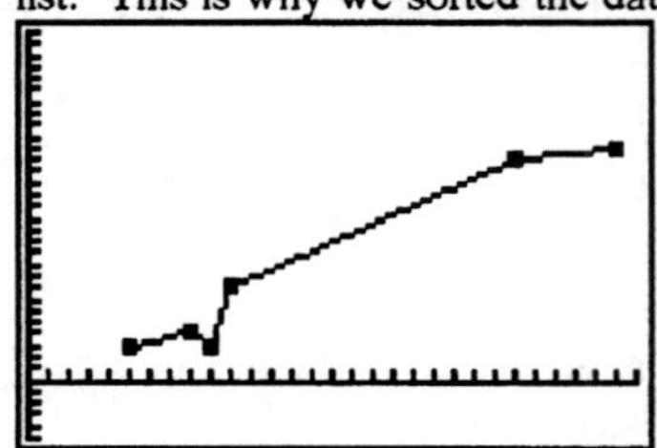

A-17 Setting the Display Mode for Numbers

Numbers can be entered into the calculator in scientific notation, a set number of significant digits, or a fixed number of decimal places.

EXAMPLE 1: Calculate $(9.35417\times 10^4)(-3.64389\times 10^{-6})$ using:
(A) scientific notation, (B) two significant digits, (C) four decimal place accuracy.

Solution (A):

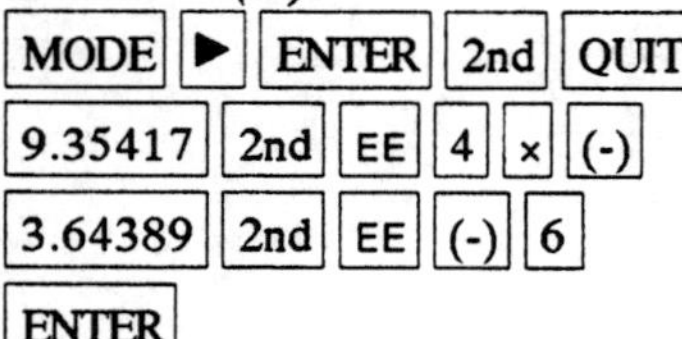

Set the calculator to scientific mode with floating decimal point.
The result displayed on the calculator is -3.408556652E-1. This means $-3.408556652\times 10^{-1}$.

Solution (B):

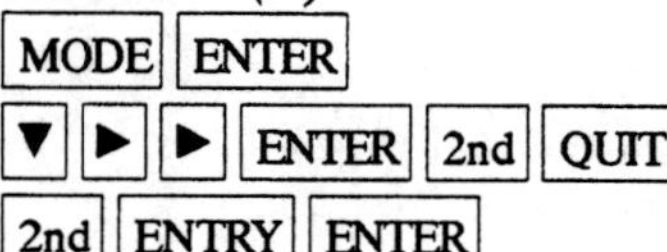

Set the calculator to scientific notation with 1 fixed decimal point instead of a floating decimal. This will give two significant digits.
Return to the Home Screen.
Use the replay key to replay the last calculation done in Solution (A). The result to two significant digits is -3.4E-1.

Solution (C):

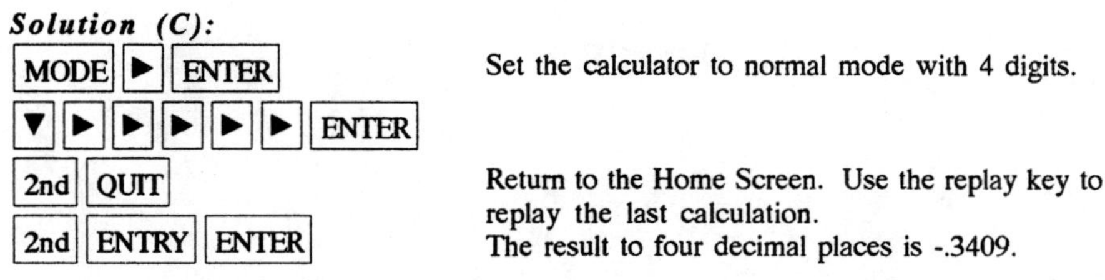

MODE ▶ ENTER — Set the calculator to normal mode with 4 digits.

▼ ▶ ▶ ▶ ▶ ▶ ENTER

2nd QUIT — Return to the Home Screen. Use the replay key to replay the last calculation.

2nd ENTRY ENTER — The result to four decimal places is -.3409.

You may now wish to return your calculator to its default settings (all items on the mode screen having the left item highlighted). The examples throughout this manual are worked using the default settings unless otherwise noted.

NOTES

NOTES

Appendix B

TI-83 Operations

B-1 Operations

Press [ON] to turn on the calculator.

Press [2nd] [+] to get the **MEMORY** screen (shown at the right).

Choose 5:Reset... by pressing [5] or use the down-arrow key [▼] to highlight 5:Reset... and press [ENTER] to get the **RESET MEMORY** screen.

Choose 2: Defaults... by pressing [2] or use the down-arrow key [▼] to highlight 2: Defaults... and press [ENTER] to get the **RESET DEFAULTS** screen.

Choose 2: Reset or use the down arrow key to highlight 2:Reset... and press [ENTER]. The screen will display Defaults set. If the screen looks blank it is because the contrast setting was also reset.

Press [2nd] and then hold the [▲] (to increase the contrast) or [▼] (to decrease the contrast) until you see the message on the screen.

Press [CLEAR] to clear the screen.

Press [2nd] [OFF] to turn the calculator off.

```
MEMORY
1: Check RAM...
2: Delete...
3: Clear Entries
4: ClrAllLists
5: Reset...
```

```
RESET MEMORY
1: All Memory...
2: Defaults...
```

```
RESET DEFAULTS
1: No
2: Reset
```

```
Defaults set
```

[2nd]

Press this key to access the operation above and to the left of a key. The operations above the key are in a gold color on the face of the calculator. Note that a flashing up arrow [↑] appears on the screen after [2nd] key is pressed.

> In this manual, the functions on the face of the calculator above a key will be referred to in square boxes just as if the function was printed on the key cap. For example, [ANS] is the function above and to the left of the [(-)] key.

ALPHA and A-LOCK

The operation above and to the right of a key is accessed by first pressing ALPHA. A flashing A is displayed as the cursor on the screen after the ALPHA key is pressed. 2nd A-LOCK locks the calculator into alpha mode until the ALPHA is pressed again.

Home Screen

Calculations are done and commands are entered on the Home Screen. You can always get to this screen (aborting any calculations in progress) by pressing

QUIT
2nd MODE. From here on, this will be referred to as 2nd QUIT in this appendix.

MODE

When you press MODE you see the screen shown below. The highlighted items are active. To change an item select it using the arrow keys and press ENTER to activate the selection.

Type of notation for display of numbers.
Number of decimal places displayed.
Type of angle measure.
Function, parametric, polar or sequence graphing.
Connected/not connected plotted points on graphs.
Graphs functions separately or all at once.
Allows a full screen or split screen to be used.

```
Normal Sci Eng
Float 0123456789
Radian Degree
Func Par Pol Seθ
Connected Dot
Sequential Simul
Rea    a+bi   re^θi
Full Horiz  G-T
```

Menus

The TI-83 Graphics calculator uses menus for selection of specific functions. The items on the menus are identified by numbers followed by a colon. There are two ways to choose menu items:

1. Press the number corresponding to the menu item.
2. Use the arrow keys to highlight the selection and then pressing ENTER.

In this manual the menu items will be referred to using the key to be pressed followed by the meaning of the menu. For example, on the ZOOM menu, 1 <ZBox> refers to the first menu item ZBox.

Editing

It is easy to correct errors when entering data into the calculator by using the arrow keys, DEL, and INS keys.

◄ or ► Moves the cursor to the left or right one position.

▲ or ▼ Moves the cursor up one line or down one line.

[DEL]	Deletes one or more characters at the cursor position.
[2nd] [INS]	Inserts one or more characters at the cursor position.
[2nd] [ENTRY]	Replays the last executed line of input.

B-2 Calculating

EXAMPLE 1: Calculate $|-4^2 - \sqrt{34}|$

Enter numbers and functions in the calculator in the same order as you would read an expression. In the examples that follow, do not press [ENTER] unless specifically instructed to do so. Enter all the keystrokes without pressing the ENTER key until [ENTER] is displayed in the example.

The subtraction sign and the negative sign are different.

The **subtraction sign** is a blue key on the lower right side of the calculator. This is used when subtracting one quantity from another.

The **negative sign** (-) is a gray key in the bottom row of keys on the calculator. This is used when entering a negative quantity.

Solution: [2nd] [QUIT] [MATH] [▶] [1] <abs(> [(-)] [4] [x^2] [-] [2nd] [√] [34] [)] [ENTER].
The result displayed is 21.83095189.

> Throughout this manual, when using menus, the number of the desired selection will be shown followed by what the selection is. Thus, [VARS] [1] <Window> means to pick the first selection (that is, Window) off the VARS menu.

B-3 Plotting Points and Drawing Line Segments

EXAMPLE 1: (A) Plot the points (-3, 6) and (1,5). (B) Draw the line segment having these points as the endpoints.

Solution (A):

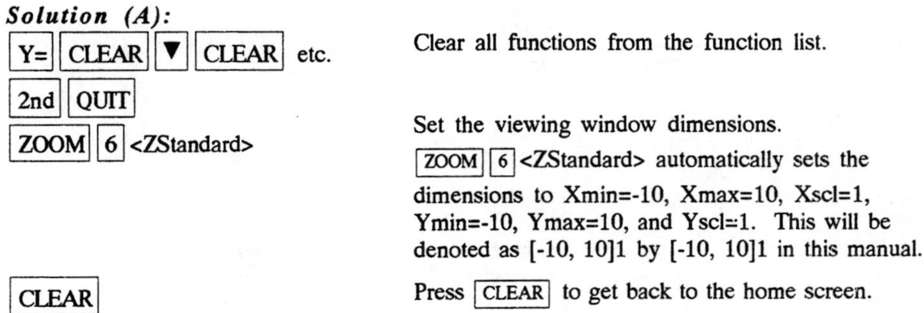

[Y=] [CLEAR] [▼] [CLEAR] etc.	Clear all functions from the function list.
[2nd] [QUIT]	
[ZOOM] [6] <ZStandard>	Set the viewing window dimensions. [ZOOM] [6] <ZStandard> automatically sets the dimensions to Xmin=-10, Xmax=10, Xscl=1, Ymin=-10, Ymax=10, and Yscl=1. This will be denoted as [-10, 10]1 by [-10, 10]1 in this manual.
[CLEAR]	Press [CLEAR] to get back to the home screen.

Keystrokes	Explanation
[2nd] [DRAW] [1] <ClrDraw> [ENTER]	Get the [DRAW] menu and clear the drawings from the calculator screen.
[2nd] [DRAW] [▶] <POINTS> [1] <PT-On(> [(-)] [3] [,] [6] [)] [ENTER]	Get the [DRAW] menu again. Choose the point on option and enter the coordinates of the point with a comma separating them. The point will be plotted.
[2nd] [QUIT]	
[2nd] [DRAW] [▶] <POINTS>	
[1] <PT-On(> [1] [,] [5] [)] [ENTER]	Enter the coordinates of the second point and plot.

Solution (B):

Keystrokes	Explanation
[2nd] [QUIT]	Return to the Home Screen.
[2nd] [DRAW] [2] <Line(>	Get the [DRAW] menu and choose the line option.
[(-)] [3] [,] [6] [,] [1] [,] [5] [)] [ENTER]	Enter the coordinates of the points separated by commas. The line segment will be drawn between the two points.

B-4 Storing and Evaluating an Expression, Lists and Tables

EXAMPLE 1: Store the expression $\frac{-b+\sqrt{b^2-4ac}}{2a}$ and evaluate it to solve $3x^2 + 4x - 5 = 0$.

Solution:

Keystrokes	Explanation
[Y=] [CLEAR]	You will see a list of function names Y1, Y2, Y3, etc. We can store ten different functions or algebraic expressions in this list.
[(] [(-)] [ALPHA] [B] [+] [2nd] [√] [ALPHA] [B] [x^2] [-] [4] [ALPHA] [A] [×] [ALPHA] [C] [)] [)] [÷] [(] [2] [ALPHA] [A] [)] [ENTER]	Then we can recall them and evaluate or graph them. We will use Y1. Press these keys to enter the expression. `(-B+√(B²-4AC))/(2A)`
[2nd] [QUIT]	Get back to the home screen.
[3] [STO▸] [ALPHA] [A] [ENTER] [4] [STO▸] [ALPHA] [B] [ENTER] [(-)] [5] [STO▸] [ALPHA] [C] [ENTER]	Store 3 as A, 4 as B and -5 as C.
[VARS] [▶] <Y-VARS> [1] <Function> [1] <Y1> [ENTER]	Recall the expression and evaluate. The result is .79 to two decimal places. So one solution to this quadratic equation is $x = .79$.

Keystrokes	Explanation
[Y=] [▶] ... [▶] [-] [ENTER]	Get the function list.
etc.	Now change the addition sign to a subtraction sign and evaluate again. The other solution is x = -2.12.

EXAMPLE 2: Given $f(x) = \sqrt[3]{x^4+2x}$ and $g(x) = 4^{2x-90}$. Find $f(x) + 3\sqrt{g(x)}$ at x = 45.2.

Solution:

Keystrokes	Explanation
[Y=] [CLEAR]	Clear the old expression for Y1 by pressing [CLEAR]
[(] [X,T,θ,*n*] [^] [4] [+] [2] [X,T,θ,*n*] [)] [^] [(] [1] [÷] [3] [)] [ENTER]	Enter $f(x)$ as Y1. (X^4+2X)^(1/3) Note that the [X,T,θ,*n*] key displays the letter X. (The letter T is displayed if parametric mode is being used, and θ is displayed if in polar graphing mode, and n is displayed if in the sequence mode.)
[CLEAR]	Clear the function stored as Y2.
[4] [^] [(] [2] [X,T,θ,*n*] [-] [90] [)] [2nd] [QUIT]	Enter $g(x)$ as Y2. 4^(2X-90)
[45.2] [STO▶] [X,T,θ,*n*] [ENTER]	Store 45.2 as X.
[VARS] [▶] <Y-VARS> [1] <Function> [1] <Y1> [+] [3] [2nd] [√]	Form $f(x) + 3\sqrt{g(x)}$ and evaluate.
[VARS] [▶] <Y-VARS> [1] <Function> [2] <Y2> [)] [ENTER]	The result is 164.97 to two decimal places.

EXAMPLE 3: Evaluate the expression $\frac{3x^2 + 5x - 8}{x^2 - 1}$ to three decimal places for x = .900, .990, .999, 1.001, 1.010, and 1.100 by: (A) Evaluating the expression for each value of x individually; and (B) using a list.

Solution (A): First store the expression in the Y= list. Then store the value for x. Recall the expression and evaluate it.

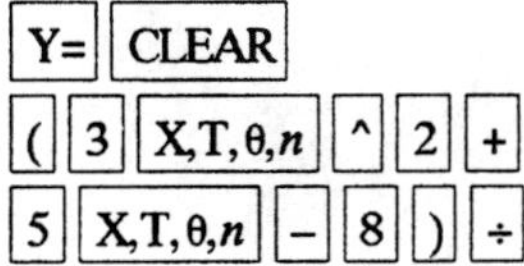

Keystrokes	Explanation
[Y=] [CLEAR]	Clear the old expression for Y1 by pressing [CLEAR].
[(] [3] [X,T,θ,*n*] [^] [2] [+] [5] [X,T,θ,*n*] [–] [8] [)] [÷]	Store the expression as Y1 and return to the home screen.

(X,T,θ,*n* ^ 2 −

1) 2nd QUIT

.9 STO▸ X,T,θ,*n* ENTER — Store the value of x.

VARS ▶ <Y-VARS> — Recall the expression and evaluate.

1 <Function> 1 <Y1> ENTER

Store the next value for x and reevaluate the expression. Repeat for the other values of x. The corresponding results are: 5.632, 5.513, 5.501, 5.499, 5.488, and 5.381.

Solution (B): Store the values of x in a list and simultaneously evaluate the expression for each value of x.

Y= CLEAR — Clear the old expression for Y1 by pressing CLEAR.

(3 X,T,θ,*n* ^ 2 + — Store the expression as Y1 and return to the home screen.

5 X,T,θ,*n* − 8) ÷

(X,T,θ,*n* ^ 2 −

1) 2nd QUIT

2nd { .9 , .99 , .999 , — Store the values of x in list L1.

1.001 , 1.01 , 1.1 2nd }

STO▸ 2nd L1 ENTER

VARS ▶ <Y-VARS> — Simultaneously calculate the value of the expression stored as Y1 for the values of x in list L1.

1 <Function> 1 <Y1> (2nd — The results are: 5.632, 5.513, 5.501, 5.499, 5.488,

L1) ENTER — and 5.381. Use the ◀ and ▶ keys to view the entire list.

EXAMPLE 4: Evaluate the expression $\frac{3x^2 + 5x - 8}{x^2 - 1}$ to three decimal places for values of x at each tenth from 0 to 1 using a table.

Solution: First store the expression in the Y= list. Set the table parameters to begin at $x = 0$ and to have an increment of .1. Get the table.

Y= CLEAR — Clear the old expression for Y1 by pressing CLEAR.

(3 X,T,θ ^ 2 + 5 X,T,θ — Store the expression as Y1 and return to the home screen.

− 8) ÷ (X,T,θ ^ 2 −

1) 2nd QUIT

2nd TblSet 0 ENTER — Set the table to begin evaluating the expression at x = 0 with a step size of .1

.1 ENTER

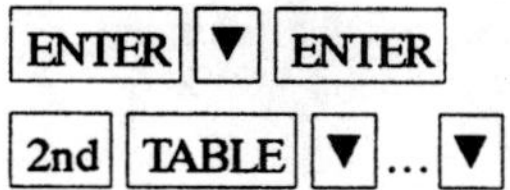

Set the calculator to automatically display values of x and Y1.

Get the table. Arrow down to see the rest of the table. The highlighted value will appear at the bottom of the table.

X	Y1	
0	8	
.1	7.5455	
.2	7.1667	
.3	6.8462	
.4	6.5714	
.5	6.3333	
.6	6.125	
X=0		

B-5 Evaluating Inequalities

EXAMPLE 1: Given the inequality $x^2 - 3t < 5x + 9t^2$. Test this inequality for $x = 1.89$ and $t = -4.52$.

Solution:

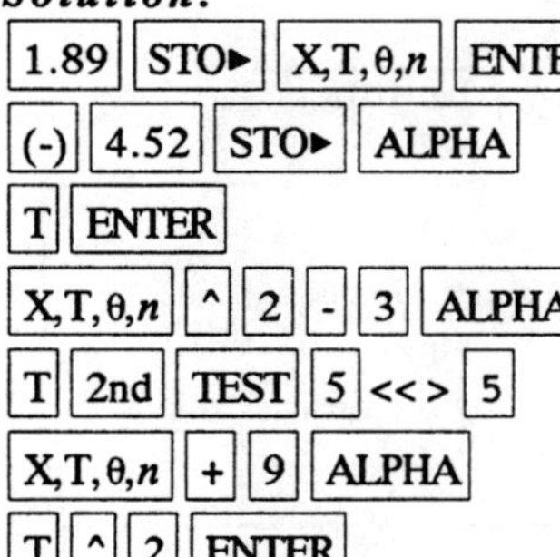

Store the value of the variables.

Enter the expression.

```
X^2-3T<5X+9T^2
```

The result is 1. This means the inequality is true for these values of the variables. If the result was a 0 then the inequality would not have been true.

B-6 Graphing and Style of Graph

EXAMPLE 1: Graph $y = 3x^2 - 6$ and $y = 3x^2 + 1$ on the same graphing screen using a solid thin line for the first function and a solid thick line for the second function.

Solution:

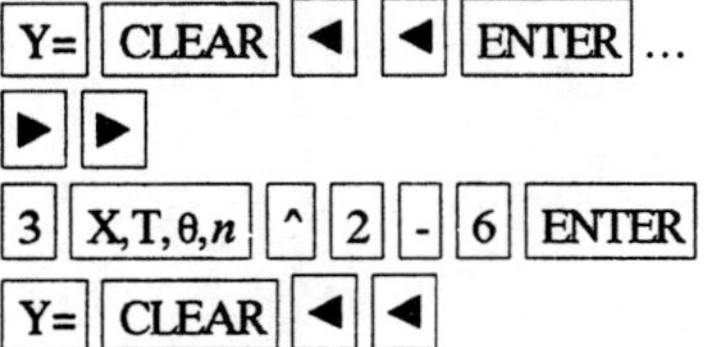

Get the function list and clear the existing function for Y1. Move the cursor to the left of the Y1. Press enter until you see the solid thin line symbol \.

Now move the cursor to the right of the equal sign and enter the first function as Y1.

Keystrokes	Explanation
[ENTER] ... [▶] [▶] [3] [X,T,θ,n] [^] [2] [+] [1] [ENTER]	Clear the second function location. Set the thick line for the graph and enter the second function as Y2. Clear or deselect all others by placing the cursor over the = sign using the left arrow and pressing enter. Now this function will not graph. Note that the = sign is no longer highlighted.
[2nd] [DRAW] [1] <ClrDraw>	Clear drawings.
[ENTER]	
[2nd] [STAT PLOT] [4] <PlotsOff>	
[ENTER]	Turn off all statistical plots. The sixth option on the ZOOM menu will automatically set the graphing screen to have $-10 < x < 10$ and $-10 < y < 10$. Also, the functions will be automatically graphed.
[ZOOM] [6] <ZStandard>	

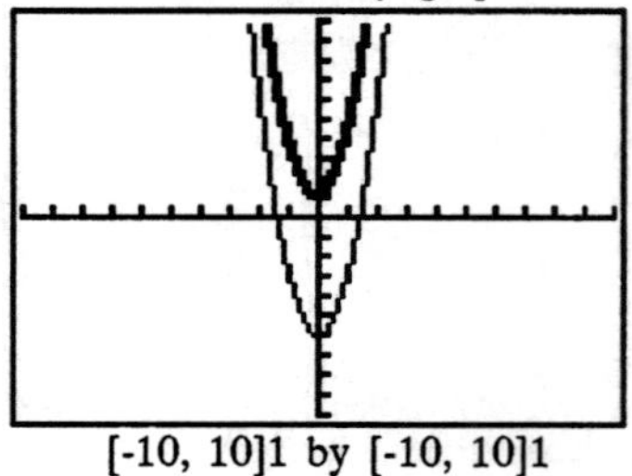

[-10, 10]1 by [-10, 10]1

> **[ZOOM] [6]** automatically sets the graphing screen limits to $-10 < x < 10$ and $-10 < y < 10$ with scale marks every unit for both the x and y axes. The graph screen limits will be denoted in this manual by the notation [-10,10]1 by [-10,10]1. The graph screen limits can be set to other values, if desired, by pressing [WINDOW] and entering the desired values, then pressing [GRAPH].

> **[6] <ZStandard>** means to choose the sixth option from the ZOOM menu.

> The function will not graph if you move the cursor over the = sign and press [ENTER]. You notice the highlighting disappears from the = sign. You can still evaluate this expression for stored values of the variables. To select the function, place the cursor on the = sign and press [ENTER] again.

EXAMPLE 2: Graph $f(x) = \begin{cases} .5x + 1 & x \le 2 \\ x^2 + 2 & x > 2 \end{cases}$.

Solution: The calculator will only graph functions for values of x for which the function is defined. So we write this piecewise-defined function in a form where the pieces are defined only for the values of x that we want. Enter the function as two separate functions with (.5 X + 1) ÷ (X ≤ 2) as Y1 and (X ^ 2 + 2) ÷ (X > 2) as Y2. The first function is defined when $x \le 2$ and the second function is defined when $x > 2$. (See Appendix Section B-5 on how to get the inequality symbols from the [TEST] menu, and see Appendix Section B-4 on how to store functions.)

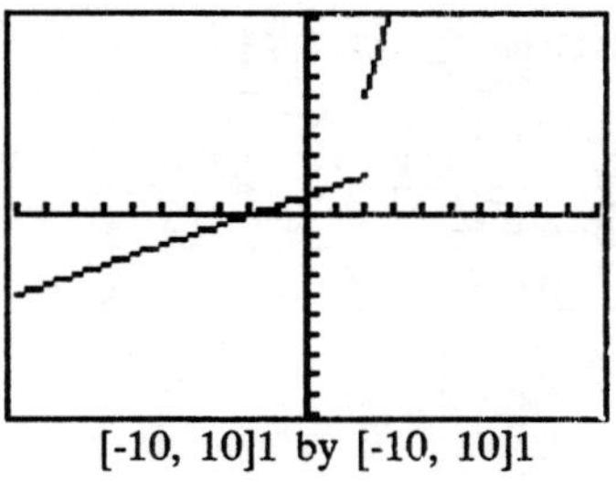

[-10, 10]1 by [-10, 10]1

B-7 TRACE, ZOOM, WINDOW, Zeroes and Intersect

EXAMPLE 3: Find the x intercept of $y = 3x^2 - 6$. Use three decimal place accuracy.

Solution: First graph $y = 3x^2 - 6$ on the standard graphing screen (see Appendix Section B-6 Example 1 above).

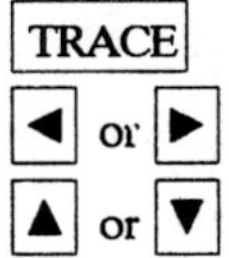

When you push [TRACE] you see a new cursor near the center of the graph screen and the coordinates of that point on the graph displayed at the bottom of the screen. Use right and left arrow keys and move the cursor along the graph.

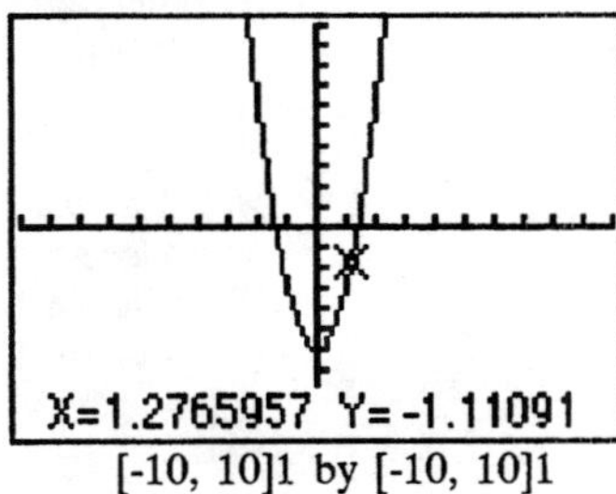

[-10, 10]1 by [-10, 10]1

There are several ways to get a closer look at the intercept:

1. Change the WINDOW dimensions.
2. Set the Zoom Factors and zoom in.
3. Use the Zoom Box feature of the calculator.
4. Use the Zero feature of the calculator.
5. Use the Intersect feature of the calculator.

METHOD 1 Set the WINDOW dimensions.

Using trace on the function graphed above we see the x intercept is near 1.5. So we shall set the graph window close to that value.

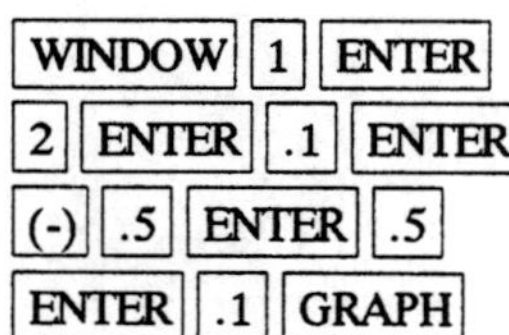

Set the WINDOW to [1, 2].1 by [-.5, .5].1. Notice we enter the minimum, maximum and scale values for the x axis, and then do the same for the y axis.

Use trace again to get a better approximation. We see that the x intercept is between 1.41 and 1.42. Set the WINDOW to [1.41, 1.42].01 by [-.01, .01].001. We now find the x intercept to be 1.414 accurate to three decimal places.

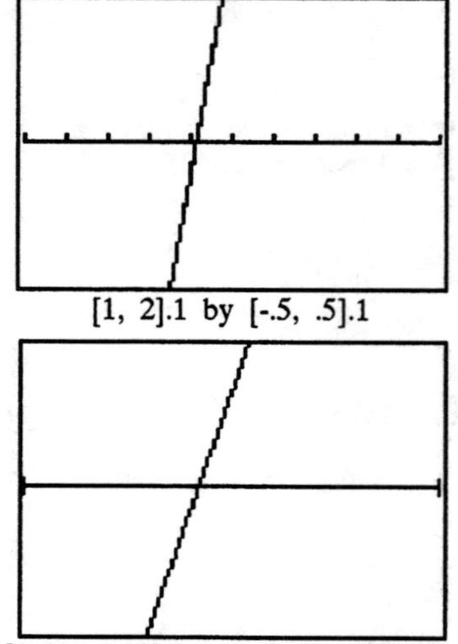

[1, 2].1 by [-.5, .5].1

[1.41, 1.42].01 by [-.01, .01].001

<u>METHOD 2 Zoom In</u>

ZOOM 6 <ZStandard>

ZOOM ▶ <MEMORY>

4 <SetFactors>

5 ENTER 5 2nd QUIT

ZOOM 2 <Zoom In>.

◀ or ▶ or ▲ or ▼

ENTER

Get the [-10, 10]1 by [-10, 10]1 display as shown in Section B-6 Example 1. Choose the MEMORY from the ZOOM menu. Choose 4 <Set Factors>.

Set the Zoom Factors to 5 and 5. After zooming in the new Xmin will be $\frac{1}{5}$ the previous Xmin.

Use the arrow keys repeatedly to get the cursor close to the x intercept, say at x = 1.4893617 y = 0. The new display will be centered where you placed the cursor.

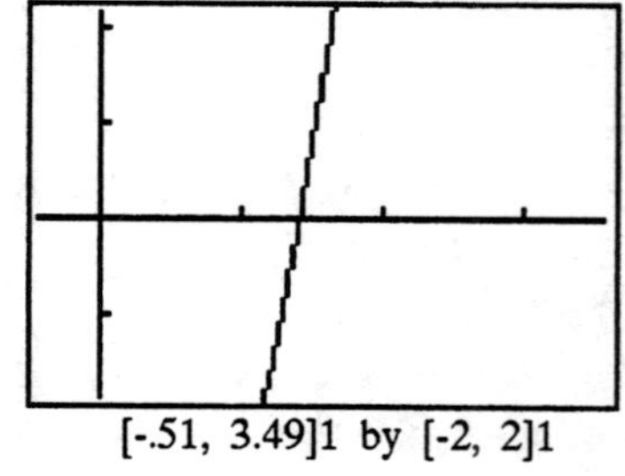

[-.51, 3.49]1 by [-2, 2]1

Now use TRACE again to get a new approximation to the x intercept.
Repeat the zooming in process as many times as necessary to get three decimal place accuracy. The x intercept is 1.414.

METHOD 3 Zoom Box

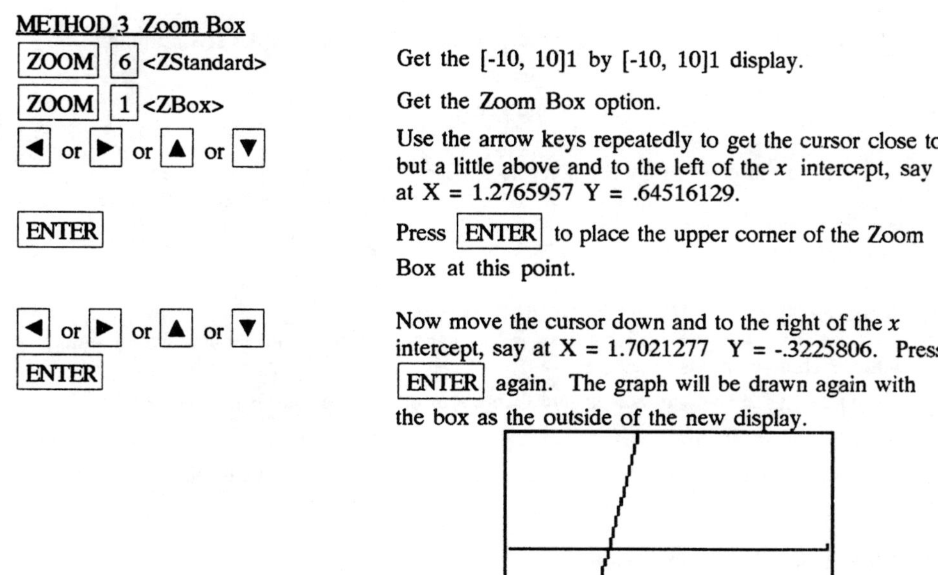

[ZOOM] [6] <ZStandard> — Get the [-10, 10]1 by [-10, 10]1 display.

[ZOOM] [1] <ZBox> — Get the Zoom Box option.

[◄] or [►] or [▲] or [▼] — Use the arrow keys repeatedly to get the cursor close to but a little above and to the left of the x intercept, say at X = 1.2765957 Y = .64516129.

[ENTER] — Press [ENTER] to place the upper corner of the Zoom Box at this point.

[◄] or [►] or [▲] or [▼] [ENTER] — Now move the cursor down and to the right of the x intercept, say at X = 1.7021277 Y = -.3225806. Press [ENTER] again. The graph will be drawn again with the box as the outside of the new display.

[1.25, 1.70]1 by [-.32, .65]1

Use trace to get an estimate of the x intercept. Use the Zoom Box feature of the calculator a second time to get an even better approximation of the x intercept. The x intercept is 1.414.

METHOD 4 zero

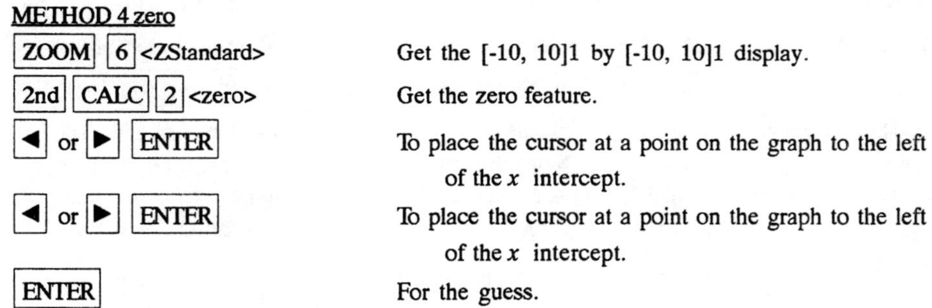

[ZOOM] [6] <ZStandard> — Get the [-10, 10]1 by [-10, 10]1 display.

[2nd] [CALC] [2] <zero> — Get the zero feature.

[◄] or [►] [ENTER] — To place the cursor at a point on the graph to the left of the x intercept.

[◄] or [►] [ENTER] — To place the cursor at a point on the graph to the left of the x intercept.

[ENTER] — For the guess.

The value of the root will appear at the bottom of the screen. The zero is (1.41, 0) rounded to two decimal places. Repeat for the other root.

METHOD 5 Intersect

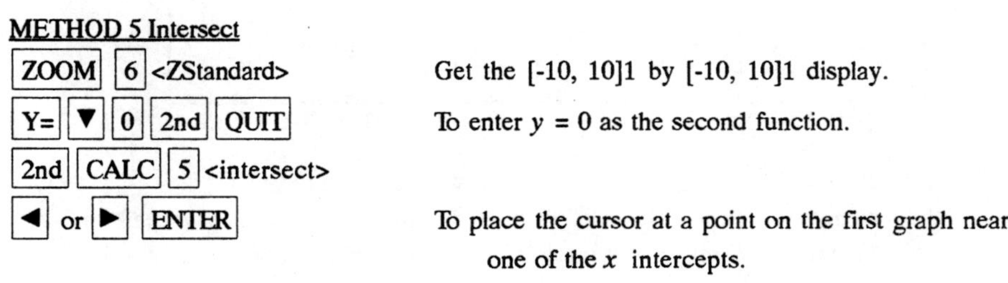

[ZOOM] [6] <ZStandard> — Get the [-10, 10]1 by [-10, 10]1 display.

[Y=] [▼] [0] [2nd] [QUIT] — To enter $y = 0$ as the second function.

[2nd] [CALC] [5] <intersect>

[◄] or [►] [ENTER] — To place the cursor at a point on the first graph near one of the x intercepts.

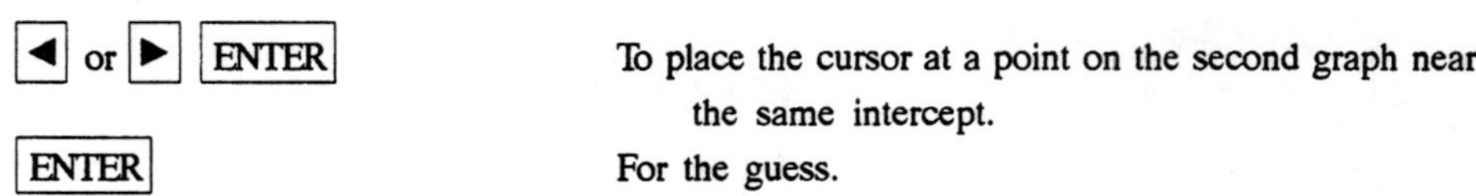

◄ or ► ENTER — To place the cursor at a point on the second graph near the same intercept.

ENTER — For the guess.

The coordinates of the point of intersection will appear at the bottom of the screen. The coordinates of the rightmost x intercept is (1.41, 0) rounded to two decimal places. Repeat for the other x intercept.

B-8 Choosing the WINDOW Variables Values

There is no one right way to determine the window variables that should be used for the graph of a function. Three methods to try are:

1. Graph using the standard screen and then zoom in or zoom out, if necessary.
2. Analyze the features of the function and then choose the viewing window dimensions.
3. Evaluate the function at several values and then choose the viewing window dimensions.

EXAMPLE 1 Graph $y = .03x^3 + .02x^2 - 5x + 10$.

Solution:

METHOD 1 Graph using the standard screen and then zoom out

Enter the function as .03 X ^ 3 + .02 X ^ 2 - 5 X + 10. Press ZOOM 6 to get the [-10,10]1 by [-10,10]1 display. (See Section B-6 of this manual.) We have too small a display to see the behavior of the graph clearly. Press ZOOM ► 4 and set the Zoom Factors to XFact = 4 and YFact = 4. Press ZOOM 3, move the cursor so it is where you want the center of the new screen to be, (we will leave it at the origin for this example) and press ENTER. The result is shown below. Note the double lines as the axes. This is because the Xscl and Yscl values do not change when zooming in. Since the scale values for x and y are small compared to the distance between Xmin and Xmax, and Ymin and Ymax, the scale marks appear to run together. You can fix this by choosing a more appropriate scale (say at 5 for Xscl and Yscl for this example).

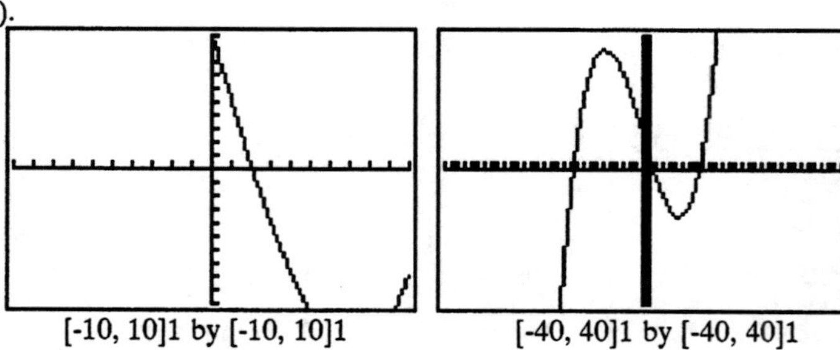

[-10, 10]1 by [-10, 10]1 [-40, 40]1 by [-40, 40]1

METHOD 2 Analyze the features of the function and then choose a WINDOW

We see that the leading coefficient is .03. This means that the vertical change in the graph will be approximately .03 unit for every time x^3 increases 1 unit. An equation that represents this is $.03x^3 = y$ or $x^3 = \frac{1}{.03}y$ or $x^3 \approx 30y$. So if y increases by 1 unit, x will increase by about $\sqrt[3]{30}$ or about 3 units. This indicates that we want the x axis three times as wide as the y axis. Hence, a good first choice for the horizontal axis is -30 to 30, and a good first choice for the vertical axis is -10 to 10. Set Xmin = -30, Xmax = 30 and Xscl = 3. We choose 3 as the Xscl because this will put 20 tick marks on the axis.

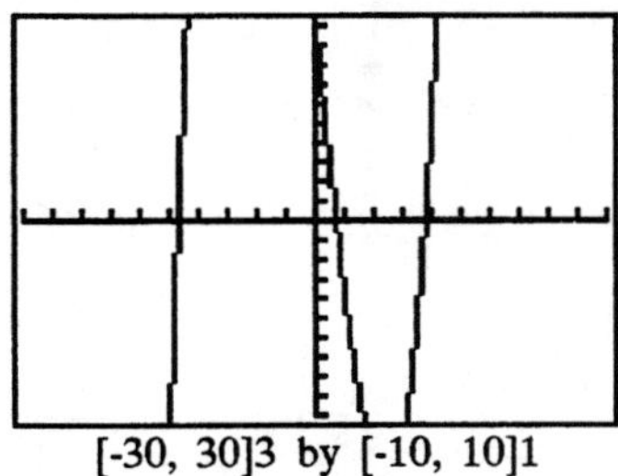

[-30, 30]3 by [-10, 10]1

The values for the vertical variables needs some adjusting since the turning points of the graph are not shown on the screen. Window variable values of [-30, 30]1 by [-40, 40]5 would show the turning points of the graph.

METHOD 3 Evaluate the function at several values and then choose the viewing window dimensions

Evaluate the function by storing a value for x and evaluating the function (see Section B-4). Repeat for each value of x. The choice of x values is arbitrary.

x	$f(x)$
-25	-321.25
-10	32
0	10
10	-8
25	366.25

Choose the Xscl and Yscl so there are about 20 marks on each axis (see METHOD 2 above). So a good first choice for the WINDOW could be [-25, 25]2 by [-300, 300]30.

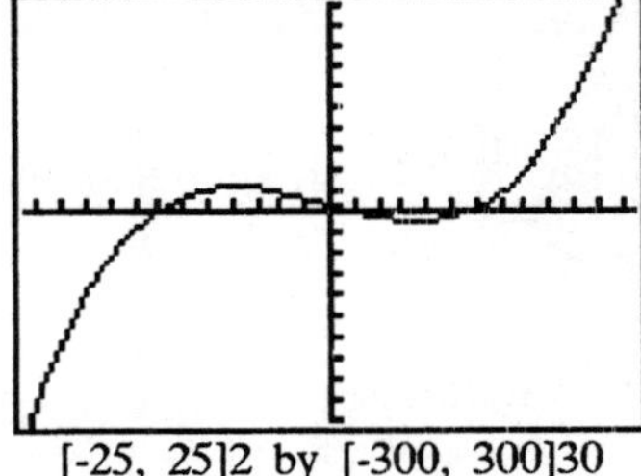

[-25, 25]2 by [-300, 300]30

B-9 Equations in One Variable

An equation in one variable can be solved graphically in two ways:

1. Algebraically move all terms to one side of the equation. Create a function $f(x)$ = (the expression). Graph the function. Zoom and trace to find the x intercepts. The set of x intercepts is the solution to the equation. (See Appendix Section B-7 of this manual.)
2. Create two functions: $f(x)$ = (the left side of the equation) and $g(x)$ = (the right side of the equation). Graph both functions. Zoom and trace to find the x value of the points of intersection of the two graphs. The set of x values is the solution to the equation.

EXAMPLE: Solve $\sqrt[3]{x + 2} = 4x^2 - 3\sqrt{x + 4}$. Use two decimal places.

Solution:

METHOD 1 Write the equation as $\sqrt[3]{x+2} - (4x^2 - 3\sqrt{x+4}) = 0$. Create the function $f(x) = \sqrt[3]{x+2} - (4x^2 - 3\sqrt{x+4})$. Enter this function into the calculator and graph.

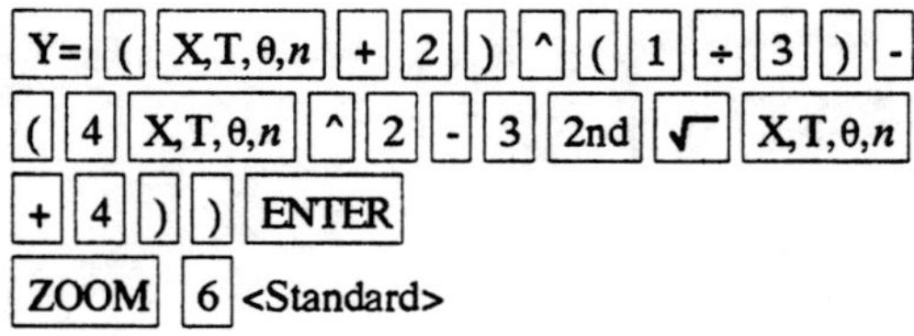

Y= (X,T,θ,n + 2) ^ (1 ÷ 3) -
(4 X,T,θ,n ^ 2 - 3 2nd √ X,T,θ,n
+ 4)) ENTER
ZOOM 6 <Standard>

Enter the function and graph.

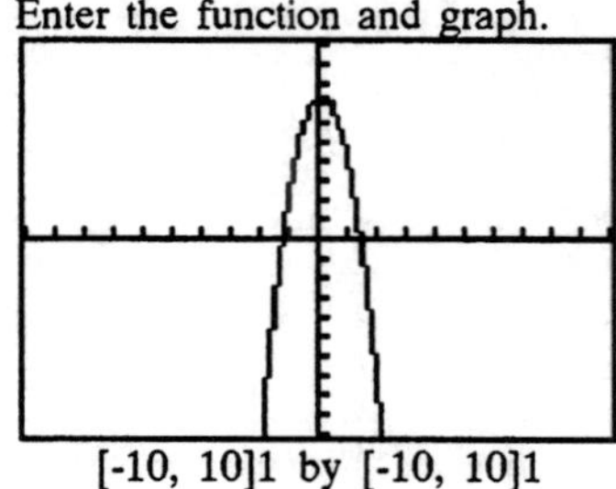

[-10, 10]1 by [-10, 10]1

We see the function crosses the x axis in two places. Trace and zoom, or use other features of the calculator, to find these x values. (See Appendix Section B-7 of this manual.) The solution to this equation is $x = -1.22$ and $x = 1.46$.

METHOD 2 Create the two functions $f(x) = \sqrt[3]{x+2}$ and $g(x) = 4x^2 - 3\sqrt{x+4}$. Graph these and find the x values of the intersection points. (See Appendix Section B-7 of this manual.)

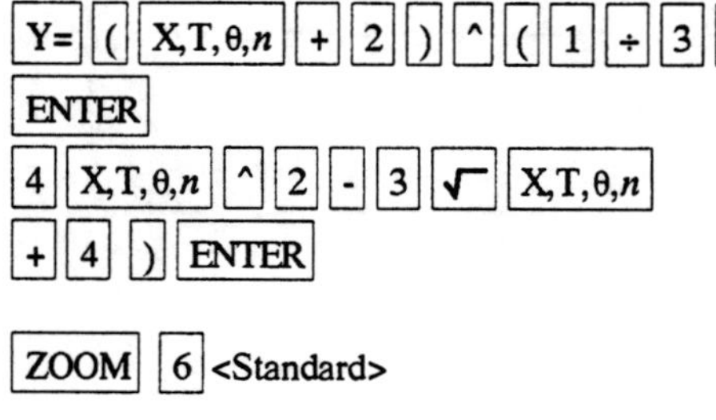

Y= (X,T,θ,n + 2) ^ (1 ÷ 3)
ENTER
4 X,T,θ,n ^ 2 - 3 √ X,T,θ,n
+ 4) ENTER

ZOOM 6 <Standard>

Enter the functions and graph.

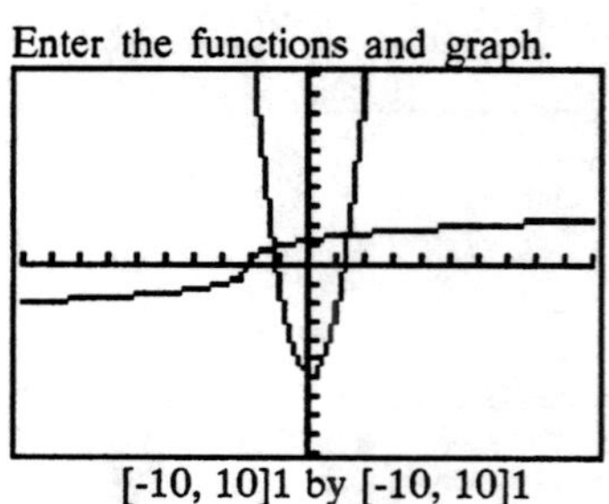

[-10, 10]1 by [-10, 10]1

We see the functions intersect in two places. Zoom and trace, or use other calculator methods, to find the coordinates of these points of intersection. (See Appendix Section B-7 of this manual.) The intersection points are approximately at (-1.22, .92) and (1.46, 1.51). So the solution to this equation is $x = -1.22$ and $x = 1.46$.

B-10 Inequalities in One Variable

An inequality in one variable can be solved graphically in two ways:

1. Algebraically move all terms to one side of the inequality and change the sign to an equal sign. Create a function $f(x)$ = (the expression). Graph the function. Zoom and trace to find the x intercepts. The set of x intercepts are the cutoff values for the solution. Now determine the intervals over which the inequality is true and write the solution. Pick a value either greater or less than the x intercept. Test the inequality using the method shown in Appendix Section B-5.

2. Create two functions: $f(x)$ = (the left side of the inequality) and $g(x)$ = (the right side of the inequality). Graph both functions. Zoom and trace to find the x value of the points of intersection of the two graphs. The set of x coordinates are the cutoff values for the solution. Now determine the intervals over which the inequality is true and write the solution. Pick a value either greater or less than the x intercept. Test the inequality using the method shown in Appendix Section B-5.

EXAMPLE: Solve $\sqrt[3]{x+2} < 4x^2 - 3\sqrt{x+4}$. Use two decimal places

Solution:

<u>METHOD 1</u> Write the inequality as: $\sqrt[3]{x+2} - (4x^2 - 3\sqrt{x+4}) < 0$. Change the sign to an equal sign: $\sqrt[3]{x+2} - (4x^2 - 3\sqrt{x+4}) = 0$. Create the function $f(x) = \sqrt[3]{x+2} - (4x^2 - 3\sqrt{x+4})$, graph, and find the x intercepts. This was done above in Appendix Section B-9 METHOD 1 of this manual. The cutoff values are $x = -1.22$ and $x = 1.46$. Observing the graph we see it is below the x axis for $x < -1.22$ and $x > 1.46$. Thus for these values of x we have $f(x)<0$ and hence

$\sqrt[3]{x+2} < 4x^2 - 3\sqrt{x+4}$. Hence the solution to this inequality can be written as $(-\infty, -1.22) \cup (1.46, \infty)$.

<u>METHOD 2</u> Create the two functions $f(x) = \sqrt[3]{x+2}$ and $g(x) = 4x^2 - 3\sqrt{x+4}$. Graph these and find the x values of the intersection points. This was done above in Appendix Section B-9 METHOD 2 of this manual. The intersection points are at (-1.22, .92) and (1.46, 1.52). Inspecting the graph we see that $f(x) < g(x)$ for $x < -1.22$ and $x > 1.46$. Hence the solution to this inequality can be written as $(-\infty, -1.22) \cup (1.46, \infty)$.

<u>B-11 Permutations, Combinations, and Random Numbers</u>

EXAMPLE 1: Find $P_{19,13}$ the number of permutations of 19 distinct objects taken 13 at a time without repetition.

Solution:

[19] [MATH] [►] [►] [►] <PRB>

[2] <nPr> [13] [ENTER]

Enter the numbers and get the built-in permutation function from the MATH menu. The result is displayed in scientific notation as 1.689515283E14.

EXAMPLE 2: Find $C_{19,13}$ the number of combinations of 19 distinct objects taken 13 at a time without repetition.

Solution:

[3] <nCr> [13] [ENTER]

Enter the numbers and get the built-in function from the MATH menu. The result is 27132.

EXAMPLE 3: Generate ten unique random numbers at least 0 and less than 50.

Solution:

Keystrokes	Explanation
[5] [STO▸] [MATH] [▸] [▸] [▸] <PRB> [1] <rand> [ENTER]	We need to store a "seed" number in the storage location RAND. This tells the random number generator where to start.
[MATH] [▸] <NUM> [3] <iPart> [50] [x] [MATH] [▸] [▸] [▸] <PRB> [1] <rand> [)] [ENTER] [ENTER]	RAND generates a number greater than 0 and less than 1. If we multiply by 50 and take the integer part, the result is a number greater than or equal to 0 and less than 50.
[ENTER] [ENTER] [ENTER] [ENTER] [ENTER] [ENTER] [ENTER] [ENTER]	The 10 random numbers are: 36 13 6 17 21 ~~21~~ 20 5 31 16 48. Note that repetitions can occur but 21 cannot be used twice.

B-12 Matrices

Given the matrices $A = \begin{bmatrix} 1 & -4 & 6 \\ 2 & -1 & 0 \\ 5 & -3 & 6 \end{bmatrix}$ and $B = \begin{bmatrix} 4 & 15 \\ -2 & 6 \\ 2 & 5 \end{bmatrix}$.

EXAMPLE 1: Find -2.5AB.

Solution:

Keystrokes	Explanation
[MATRX] [▶] [▶] <EDIT> [1] <[A]>	Choose matrix A and enter the dimensions.
[3] [ENTER] [3] [ENTER]	
[1] [ENTER] [(-)] [4] [ENTER] [6] [ENTER] [2] [ENTER] [(-)] [1] [ENTER] [0] [ENTER] [5] [ENTER] [(-)] [3] [ENTER] [6] [ENTER]	Enter the matrix elements.
[MATRX] [▶] [▶] <EDIT> [2] <[B]>	Choose matrix B and enter the dimensions.
[3] [ENTER] [2] [ENTER]	
[4] [ENTER] [15] [ENTER] [(-)] [2] [ENTER] [6] [ENTER] [2] [ENTER] [5]	Enter the matrix elements.
[2nd] [QUIT]	Return to the Home Screen. Note that [A] is the matrix A. This is different than A which represents the variable A.

[(-)] [2.5] [MATRX] [1] <[A]> [x] [MATRX] [2] <[B]> [ENTER]

Enter the matrix operations.

The result is $\begin{bmatrix} -60 & -52.5 \\ -25 & -60 \\ -95 & -217.5 \end{bmatrix}$.

EXAMPLE 2: Find A^{-1} using matrix A from Example 1.

Solution:

[MATRX] [1] <[A]> [x^{-1}] [ENTER]

The operations are done from the Home Screen. The result rounded to three decimal places is:

$$A = \begin{bmatrix} -.167 & .167 & .167 \\ -.333 & -.667 & .333 \\ -.028 & -.472 & .194 \end{bmatrix}.$$

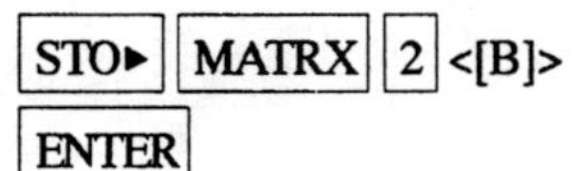

To inspect the result use the arrow keys to see the elements. The mode can be set to three decimal places. See Appendix Section B-17 of this manual.

EXAMPLE 3: Solve the system of equations using the inverse of the coefficient matrix. Round answers to two decimal places.

$$\begin{aligned} x - 4y + 6z &= 2.5 \\ 2x - y &= 3.2 \\ 5x - 3y + 6z &= -3.4 \end{aligned}$$

Solution: Note that the coefficient matrix is matrix A from Section B-12 Example 1 of this manual. The method of finding the inverse of a matrix was shown above in Section B-12 Example 2. Enter the constants as a 3×1 matrix B and find $A^{-1}B$.

[MATRX] [▶] [▶] <EDIT> [2] <[B]> — Enter the matrix dimensions.

[3] [ENTER] [1] [ENTER]

[2.5] [ENTER] — Enter the matrix elements.

[3.2] [ENTER] [(-)] [3.4]

[2nd] [QUIT] — Return to the Home Screen.

[MATRX] [1] <[A]> [x^{-1}] [MATRX] [2] <[B]> [ENTER] — Perform the matrix operations.

The result is $\begin{bmatrix} -.45 \\ -4.10 \\ -2.24 \end{bmatrix}$. The solution to the system of equations to two decimal places is $x = -.45$, $y = -4.10$, $z = -2.24$.

EXAMPLE 4: Find the solution to the system of Example 3 above using Gauss-Jordan reduction. In other words, find the reduced form of the matrix $\begin{bmatrix} 1 & -4 & 6 & 2.5 \\ 2 & -1 & 0 & 3.2 \\ 5 & -3 & 6 & -3.4 \end{bmatrix}$.

Two different methods can be used:

1. Use the matrix math row operation features.
2. Use the reduced row-echelon form matrix math feature.

Each method is described below. First we need to enter the matrix elements into the calculator.

Keystrokes	Comment
[MATRX] [▶] [▶] <EDIT> [1] <[A]>	Name the matrix A. Enter the matrix dimensions.
[3] [ENTER] [4] [ENTER]	
[1] [ENTER] [(-)] [4] [ENTER]	Enter the matrix elements row by row.
[6] [ENTER] [2.5] [ENTER]	
[2] [ENTER] [(-)] [1] [ENTER]	
[0] [ENTER] [3.2] [ENTER]	
[5] [ENTER] [(-)] [3] [ENTER]	
[6] [ENTER] [(-)] [3.4]	
[2nd] [QUIT]	Return to the home screen.

After a calculation, some numbers may look like decimals. However, upon storing, recalling, and examining the matrix using the arrow keys you will find they are decimals that are very close to zero. This situation can be eliminated by setting the number of decimal places to a number, say 3 (see Section B-3 of this manual).

<u>METHOD 1</u>

The reduced form of the matrix can be found step by step using the matrix math commands in the calculator.

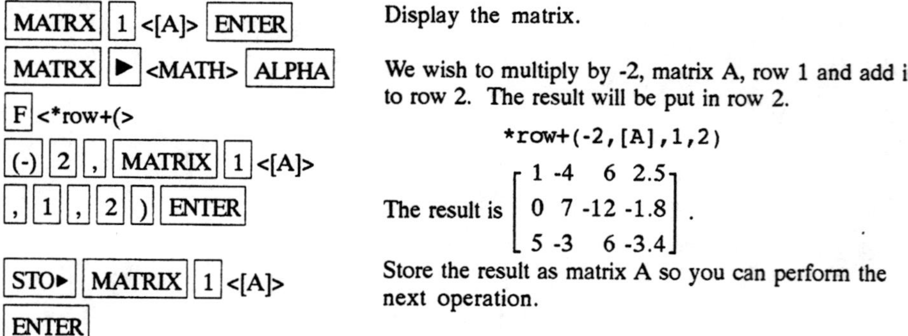

Keystrokes	Comment
[MATRX] [1] <[A]> [ENTER]	Display the matrix.
[MATRX] [▶] <MATH> [ALPHA] [F] <*row+(>	We wish to multiply by -2, matrix A, row 1 and add it to row 2. The result will be put in row 2.
[(-)] [2] [,] [MATRIX] [1] <[A]> [,] [1] [,] [2] [)] [ENTER]	`*row+(-2,[A],1,2)` The result is $\begin{bmatrix} 1 & -4 & 6 & 2.5 \\ 0 & 7 & -12 & -1.8 \\ 5 & -3 & 6 & -3.4 \end{bmatrix}$.
[STO▶] [MATRIX] [1] <[A]> [ENTER]	Store the result as matrix A so you can perform the next operation.

Keystrokes	Explanation
MATRX ▶ <MATH> ALPHA F <*row+(> (-) 5 , MATRIX 1 <[A]> , 1 , 3) ENTER	Now multiply by -5, matrix A, row 1 and add it to row 3. `*row+(-5,[A],1,3)` The result is $\begin{bmatrix} 1 & -4 & 6 & 2.5 \\ 0 & 7 & -12 & -1.8 \\ 0 & 17 & -24 & -15.9 \end{bmatrix}$.
STO▸ MATRIX 1 <[A]> ENTER	Store the result as matrix A.
MATRX ▶ <MATH> E <*row(> 1 ÷ 7 , MATRIX 1 <[A]> , 2) ENTER	Multiply by 1/7, matrix A, row 2. `*row(1÷7,[A],2)` The result is $\begin{bmatrix} 1 & -4 & 6 & 2.5 \\ 0 & 1 & -1.71 & -.26 \\ 0 & 17 & -24 & -15.9 \end{bmatrix}$ rounded to two decimal places.
STO▸ MATRIX 1 <[A]> ENTER	Store the result as matrix A.
	Continue in this manner until you get $\begin{bmatrix} 1 & 0 & 0 & -.45 \\ 0 & 1 & 0 & -4.10 \\ 0 & 0 & 1 & -2.24 \end{bmatrix}$. The solution to the system of equations is $x = -.45$, $y = -4.10$, $z = -2.24$.

> NOTE: To swap rows of a matrix use ALPHA C <rowSwap(> from the MATRX MATH menu. The command rowSwap([A],2,3) will swap rows 2 and 3 in matrix A.
>
> Note: To add one row to another use ALPHA D <row+(> from the MATRX MATH menu. The command row+([A],2,3) will add row 2 to row 3 in matrix A.

METHOD 2

The reduced row-echelon form can be found using the rref command in the calculator.

MATRX ▶ <MATH> ALPHA B <rref> MATRX 1 <[A])

B-13 Logarithmic, Exponential, and Hyperbolic Functions

EXAMPLE 1: Graph $f(x) = 10^{3.2x}$ for $-.5 < x < .5$.

Solution: Set the WINDOW to [-.5,.5].1 by [-10,10]1. (See Appendix C Section B-7 Example 1 Method 1.)

[Y=] [10] [^] [(] [3.2] [X,T,θ,n] [)]
[GRAPH]

Enter the function and graph.

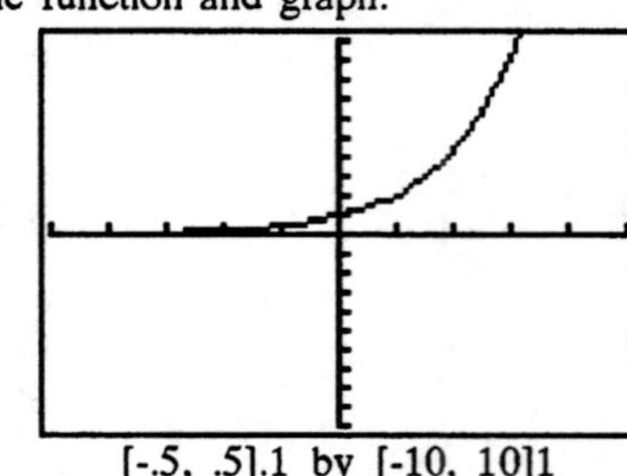

[-.5, .5].1 by [-10, 10]1

EXAMPLE 2: Graph $y = \dfrac{e^x + e^{-x}}{2}$.

Solution: This could be entered as [(] [2nd] [e^x] [X,T,θ,n] [+] [2nd] [e^x] [(-)] [X,T,θ,n] [)] [÷] [2]. Recall that [e^x] represents the built-in exponential function above the [LN] key. However this is the hyperbolic cosine function. The following uses this built-in function.

[Y=] [CLEAR]
[2nd] [CATALOG] [▼] ... [▼] [ENTER]
[X,T,θ,n] [)]
[ZOOM] [6] <ZStandard>

Clear the existing functions from the Y= list.
Get the catalog of functions, arrow down until the cursor is at the cosh function, and press ENTER.

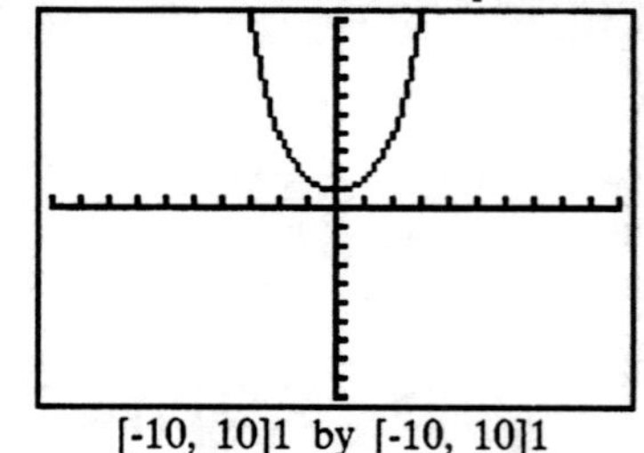

[-10, 10]1 by [-10, 10]1

EXAMPLE 3: Solve $\ln(x^2 + x) = 3x + 2$

Solution:

[Y=] [CLEAR] [LN] [X,T,θ,n]
[^] [2] [+] [X,T,θ,n] [)] [ENTER]

Enter the function $f(x) = \ln(x^2 + x)$ as Y_1 in the function list.

[CLEAR] [3] [X,T,θ,n]
[+] [2] [ENTER]

Enter the function $g(x) = 3x + 2$ as Y_2 in the function list.

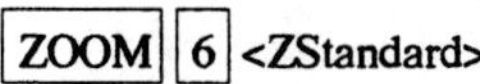

Graph them on the same graphing screen. Trace and Zoom, or use other calculator methods, to find the *x* value of the point of intersection. (See Appendix Section B-7 of this manual.) This will be the solution to the equation.

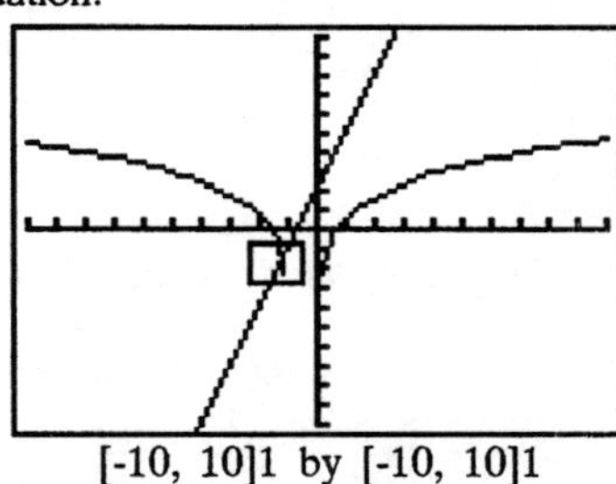

[-10, 10]1 by [-10, 10]1

The solution is -1.18, rounded to two decimal places.

B-14 Trigonometric Functions

EXAMPLE 1: Evaluate $f(x) = \cos x + \csc x$ at (A) 53°45'18" and (B) 78.52 radians.

Solution: (A)

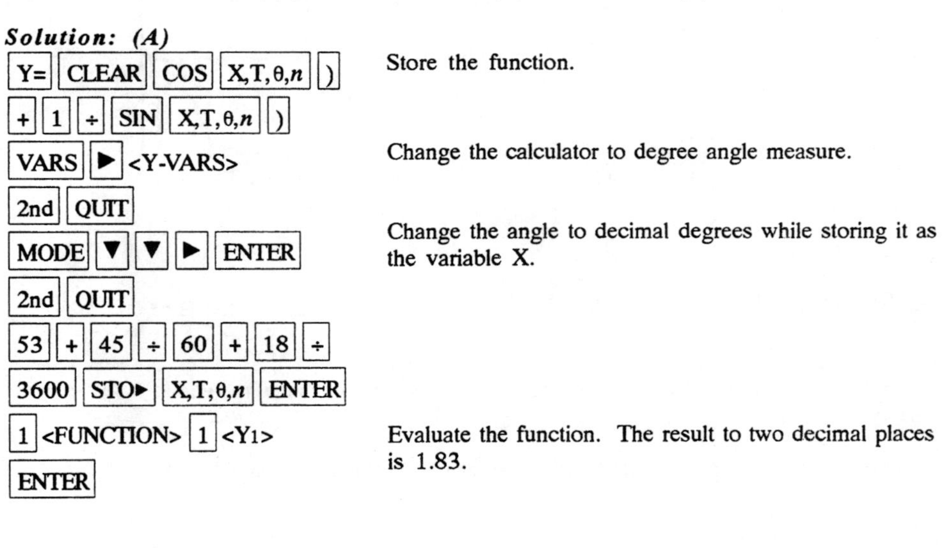

Store the function.

Change the calculator to degree angle measure.

Change the angle to decimal degrees while storing it as the variable X.

Evaluate the function. The result to two decimal places is 1.83.

Solution: (B)

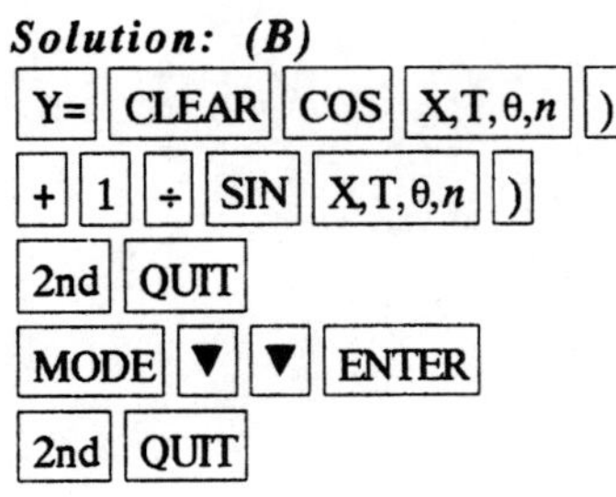

Store the function.

Change the calculator to radian angle measure. Since the function was stored in Part (A) above, we need only to store the angle measure and evaluate the stored function.

Keystrokes	Explanation
[78.52] [STO▶] [X,T,θ,n] [ENTER]	Store 78.52 as X.
[VARS] [▶] <Y-VARS> [1] <FUNCTION> [1] <Y1> [ENTER]	Evaluate the function. The result to two decimal places is 49.47.

EXAMPLE 2: Graph $f(x) = \cos x + \csc x$.

Solution: The function was entered in the calculator in Part (A) above.

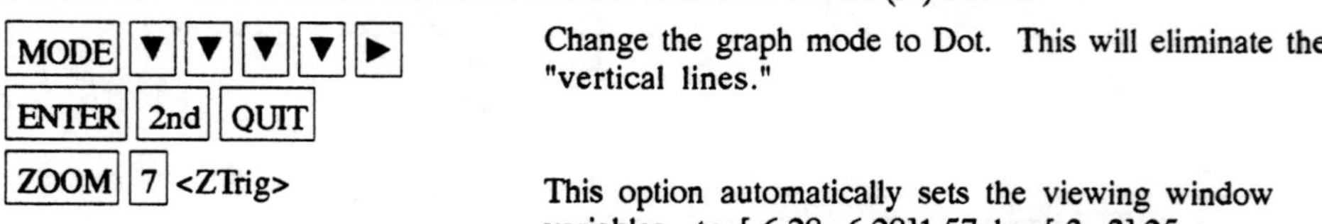

Keystrokes	Explanation
[MODE] [▼] [▼] [▼] [▼] [▶] [ENTER] [2nd] [QUIT]	Change the graph mode to Dot. This will eliminate the "vertical lines."
[ZOOM] [7] <ZTrig>	This option automatically sets the viewing window variables to [-6.28, 6.28]1.57 by [-3, 3].25.

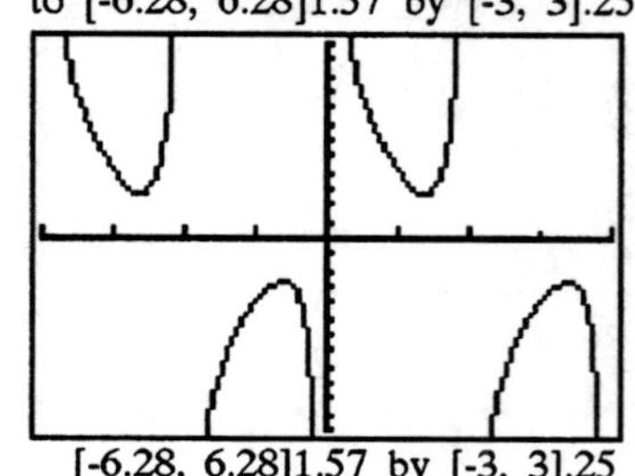

[-6.28, 6.28]1.57 by [-3, 3].25

B-15 Numeric Differentiation and Integration, and Tangent Lines

EXAMPLE 1: (A) Find the equation of the tangent line to $f(x) = \sqrt{x^5 - \sqrt[3]{x}}$ at $x = 3.54$. Use two decimal place accuracy. (B) Graph the function and the tangent line at $x = 3.54$ using [0, 6]1 by 0, 50]2.

Solution (A): The function and its derivative can be evaluated at this value of x.

Keystrokes	Explanation
[Y=] [CLEAR] [2nd] [√] [X,T,θ,n] [^] [5] [-] [X,T,θ,n] [^] [(] [1] [÷] [3] [)] [)]	Enter the function and store it as Y1.
[2nd] [QUIT] [3.54] [STO▶] [X,T,θ,n] [ENTER]	Get the Home Screen and store the value of x.

VARS ▶ <Y-VARS>
1 <FUNCTION> 1 <Y1>
ENTER

Evaluate the function. $f(3.54) = 23.55$ to two decimal places.

MATH 8 <nDeriv(>
VARS ▶ <Y-VARS>
1 <FUNCTION> 1 <Y1>
, X,T,θ,*n* , 3.54 , .0001)
ENTER

Get the numeric derivative built-in function.

```
nDeriv(Y1,X,3.54,.0001)
```

This will evaluate the derivative with respect to x of the function stored in Y1 at $x = 3.54$ with .0001 tolerance. If the tolerance is not entered, the calculator will use .001.

The result is 16.67.

The equation of the tangent line is $y - 23.55 = 16.67(x - 3.54)$.

Solution (B): The tangent line can be drawn from the Home Screen. Set the WINDOW dimensions to [0, 5]1 by [0, 50]2.

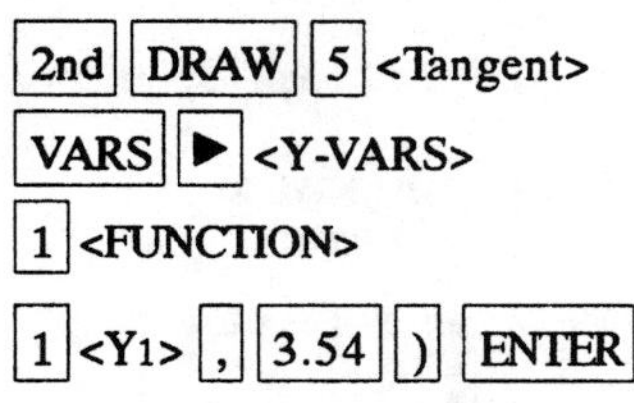

2nd DRAW 5 <Tangent>
VARS ▶ <Y-VARS>
1 <FUNCTION>
1 <Y1> , 3.54) ENTER

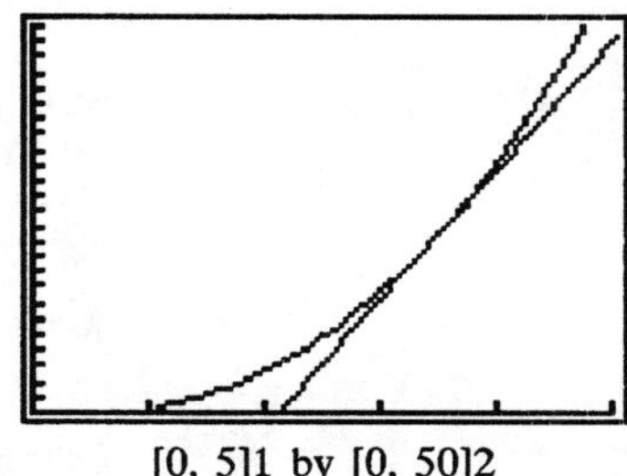

[0, 5]1 by [0, 50]2

EXAMPLE 2: Find $\int_{1.32}^{2.58} \sqrt{x^5 - \sqrt[3]{x}}\, dx$. Use two decimal place accuracy.

Solution:

Y= CLEAR 2nd √ X,T,θ,*n* ^
5 - X,T,θ,*n* ^ (1 ÷ 3))

Enter the function and store it as Y1.

2nd QUIT
MATH 9 <fnInt(>
VARS ▶ <Y-VARS> 1 <FUNCTION>
1 <Y1> , X,T,θ,*n* , 1.32 , 2.58
, .00001) ENTER

Get the numeric integration function built into the calculator function:

```
fnInt(Y1,X,1.32,2.58)
```

This will evaluate the definite integral. The result is 6.95. The tolerance in calculating the definite integral is .00001.

B-16 Statistics

EXAMPLE 1: Given the following data find: (A) the mean, (B) the sample standard deviation, (C) an histogram with first class beginning at 0 and width 4, and (D) repeat Parts A and B after deleting the data value 89 from the list. How did eliminating this outlier effect the mean and standard deviation?

23	56	73	12	89	52	45	19	40	57	39
38	18	21	15	37	42	51	38	54	60	

Use one decimal place in answers.

Solution (A) and (B):

Keystrokes	Explanation
[STAT] [4] <ClrList> [2nd] [L1] [ENTER]	Clear the list named L1.
[STAT] [1] <Edit>	Get the data entry menu.
[23] [ENTER] [56] [ENTER] Etc.	Enter the data in the L1 list. Enter all data. There are 21 values. If you make a mistake, use the arrow keys to highlight the number and type it in again. To remove a data value, use the arrow keys to highlight number and press [DEL].
[2nd] [QUIT]	Exit the data entry screen.
[STAT] [►] <CALC> [1] <1-Var Stats> [2nd] [L1] [ENTER]	Get the statistics menu again. Press [1] to get the one variable statistics. Since each piece of data was entered individually, we need only to list L1 after 1-Var Stats. The calculator will use 1 as the frequency for each number in the L1 list. The mean is 41.86 and the sample standard deviation is 19.82.

Solution (C):

Keystrokes	Explanation
[WINDOW] [0] [ENTER] [100] [ENTER] [4] [ENTER] [(-)] [5] [ENTER] [5] [ENTER] [1]	The smallest value is 12 so let xMin=0. The greatest value is 89. Let us set Xmax=100. Let Xscl=4 since this is the class width. Let Ymin=-5 to get some room below the histogram. We do not know the frequencies yet so let Ymax=5 for a first try. Let yScl=1.
[Y=] [CLEAR] ...	Clear all functions on the function list.
[2nd] [DRAW] [1] <ClrDraw> [ENTER]	Clear all drawings from the display.

2nd STAT PLOT 1 <Plot1...OFF>	Select Plot1
ENTER	Set Plot1 to ON.
▼ <Type:> ▶ ▶ < ⌊ > ENTER	Select Type as Histogram.
▼ <Xlist:> ENTER	Select Xlist as L1 and Freq as 1.
▼ ENTER	
GRAPH	Graph the histogram.

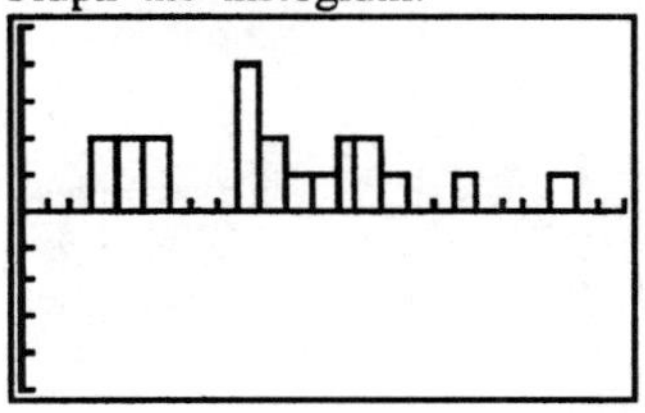

Solution (D):

STAT 1 ▼ ▼ ... ▼ ▼ DEL	Get the data list and arrow down to the 89. Press DEL.
STAT ▶ <CALC> 1 <1-Var Stats> 2nd L1 ENTER	Get the statistics menu and the one variable sample statistics. The mean is now 39.5 and the sample standard deviation is 17.05. Notice how much influence an outlier has!

EXAMPLE 2: Given the ungrouped frequency data, shown at the right: (A) Find the mean. (B) Find the sample standard deviation. (C) Draw a boxplot.

x	f
2.4	4
3.5	6
4.9	8
5.3	5
6.8	2

Solution (A) & (B):

STAT 4 <ClrList> 2nd L1 , 2nd L2 ENTER	Clear the lists named L1 and L2.
STAT 1 <Edit>	Get the data entry menu.
2.4 ENTER 3.5 ENTER Etc.	Enter the data in the L1 list. Enter all data. If you make a mistake, use the arrow keys to highlight the number and type it in again. To remove a data value, use the arrow keys to highlight number and press DEL.

Keystrokes	Explanation
► 4 ENTER 6 ENTER Etc.	Enter the frequencies in the L2 list.
2nd QUIT	Exit the data entry screen.
STAT ► <CALC> 1 <1-Var Stats>	Get the statistics menu again. Press 1 to get the
2nd L1 , 2nd L2 ENTER	one variable statistics. The calculator will use the L1 list for the values of x and the L2 as the frequency for each number in the L1 list. The mean is 4.40 and the sample standard deviation is 1.27, rounded to two decimal places.

Solution (C):

Keystrokes	Explanation
WINDOW ENTER 0 ENTER	The smallest value is 2.4 so let Xmin=0.
10 ENTER 1 ENTER	The greatest value is 6.8 so let Xmax=10. Let Xscl=1. Let Ymin=-1 to get some room below the graph. Let Ymax=4 and Yscl=1. Leave Xres at 1.
(-) 1 ENTER	
4 ENTER 1 ENTER 1	
Y= CLEAR ...	Clear all functions on the function list.
2nd DRAW 1 <ClrDraw> ENTER	Clear all drawings from the display.
2nd STAT PLOT 1 <Plot1...OFF>	Select Plot1
ENTER	Set Plot1 to ON.
▼ <Type:>	Select Type as Boxplot.
► ► ► ► < > ENTER	
▼ <Xlist:> 2nd L1 ▼ 2nd L2	Select Xlist as L1 and Freq as L2.
GRAPH	Graph the boxplot.

EXAMPLE 2: Given the following bivariate data: (A) Find the correlation coefficient. (B) Find the equation of the least squares line. (C) Graph the least squares line and the scatter plot on the same graph. (D) Plot the scatter diagram without the least squares line. (E) Sort the data on x and join the points using straight line segments.

x	5	8	10	29	9	24
y	3	4	8	20	3	19

Solution (A):

Keystrokes	Explanation
[STAT] [4] <ClrList> [2nd] [L1] [,] [2nd] [L2] [ENTER]	Clear list L1 and list L2.
[STAT] [1] <Edit>	Get the lists.
[5] [ENTER] [8] [ENTER] [10] [ENTER] [29] [ENTER] [9] [ENTER] [24] [ENTER]	Enter the data in list L1.
[▶] [3] [ENTER] [4] [ENTER] [8] [ENTER] [20] [ENTER] [3] [ENTER] [19] [ENTER]	Move to list L2. Enter the data in list L2. The data for y in L2 must correspond pairwise to the data for x in L1.
[STAT] [▶] <CALC> [8] <LinReg(a+bx)> [2nd] [L1] [,] [2nd] [L2] [ENTER]	Get the linear regression coefficients using L1 as the x variable and L2 as the y variable in the calculations.
[STAT] [▶] [▶] <Tests> [ALPHA] [E] <LinRegTTest> [2nd] [L1] [ENTER] [2nd] [L2] [ENTER] [ALPHA] [1] [ENTER] [▼] [ENTER] [ENTER] [▼] [▼] [▼] [▼]	This gives you many pieces of information. The information we want is: a = -1.72, b = .79, r = .98. The linear correlation coefficient is r = .98. The equation of the line is $\hat{y}$ = -1.72 + .79x.

Solution (B):

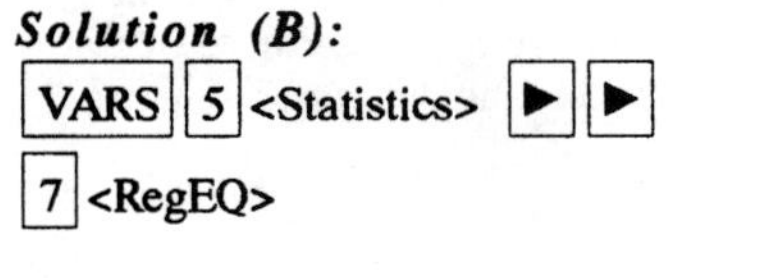

Keystrokes	Explanation
[VARS] [5] <Statistics> [▶] [▶] [7] <RegEQ>	The least squares line equation is displayed. It is $\hat{y}$ = -1.72 + .79x with the coefficients rounded to two decimal places.

Solution (C):

Keystrokes	Explanation
[WINDOW] [0] [ENTER] [30] [ENTER] [1] [ENTER] [(-)] [5] [ENTER] [30] [ENTER] [1]	Set the graph WINDOW. Observe the data to determine the WINDOW values. Set the y value lower than needed so if you trace, the coordinates at the bottom of the screen do not cover up the graph.
[Y=] [CLEAR] ...	Clear all functions on the function list.
[2nd] [DRAW] [1] <ClrDraw> [ENTER]	Clear all drawings from the display.

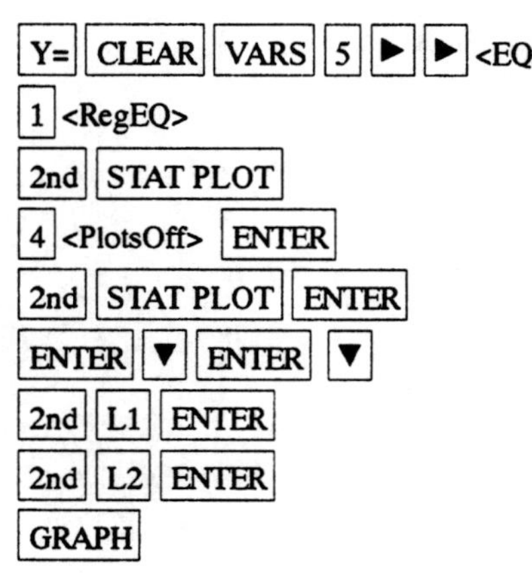

Clear or deselect any existing functions in the list. Clear Y1.

The least squares line equation will be put into the function list so you can graph it.

Set up the calculator to get the scatter plot with the line.

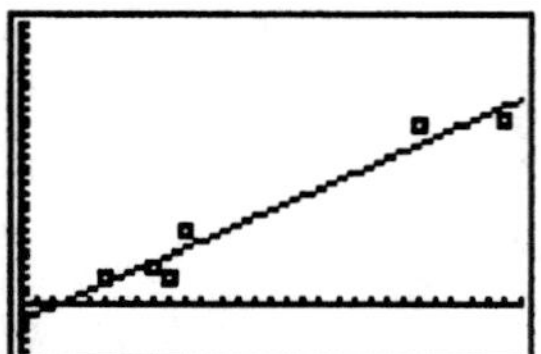

Solution (D):

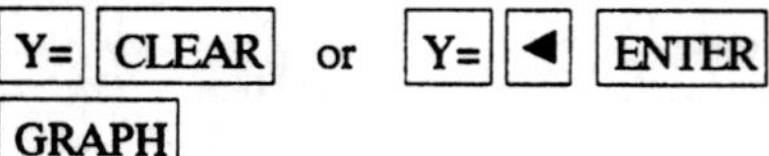

Clear the functions from the list or deselect them so they won't graph.
Plot the points on the graph.

Solution (E):

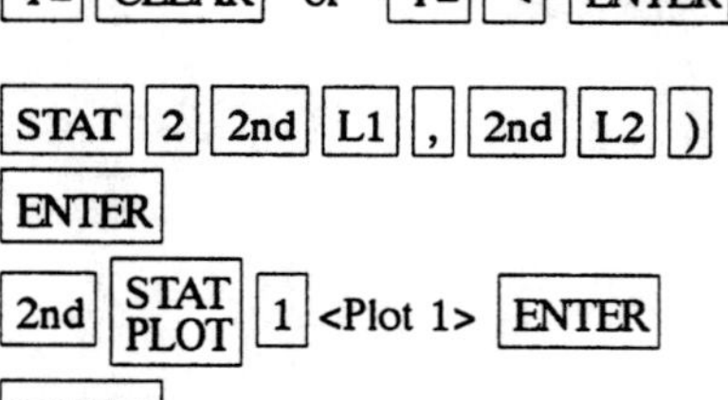

▼ ► ENTER ▼ 2nd L1 ENTER

2nd L2

GRAPH

Clear the functions from the list or deselect them so they won't graph.

Sort the data on the x variable. The list L1 will be sorted in ascending order. The list L2 will sort along with L1 as pairs.
Set the calculator to draw line segments joining the points.
This will join the points together with straight line segments in the order they appear in the data list. This is why we sorted the data on x first.

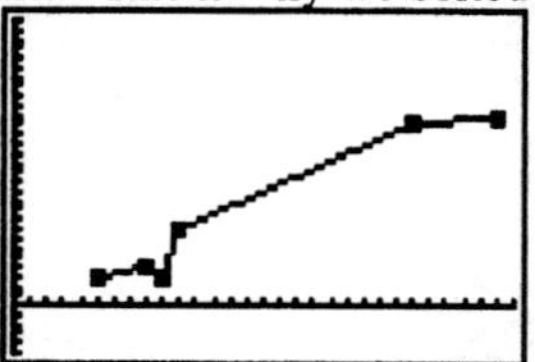

B-17 Setting the Display Mode for Numbers

Numbers can be entered into the calculator in scientific notation, a set number of significant digits, or a fixed number of decimal places.

EXAMPLE 1: Calculate $(9.35417\times 10^4)(-3.64389\times10^{-6})$ using:
(A) scientific notation (B) two significant digits (C) four decimal place accuracy.

Solution (A):

MODE ▶ ENTER 2nd QUIT — Set the calculator to scientific mode with a floating decimal point.

9.35417 2nd EE 4 × (-)

3.64389 2nd EE (-) 6

ENTER — The result displayed on the calculator is -3.408556652E-1. This means $-3.408556652 \times 10^{-1}$.

Solution (B):

MODE ▶ ENTER

▼ ▶ ▶

ENTER — Set the calculator to scientific notation with 1 fixed decimal point instead of a floating decimal. This will give two significant digits.

2nd QUIT — Return to the Home Screen.

2nd ENTRY ENTER — Use the replay key to replay the last calculation done in Solution (A). The result to two significant digits is -3.4E-1.

Solution (C):

MODE ENTER

▼ ▶ ▶ ▶ ▶ ▶ ENTER — Set the calculator to normal mode with 4 digits.

2nd QUIT

2nd ENTRY ENTER — Return to the Home Screen. Use the replay key to replay the last calculation. The result to four decimal places is -.3409.

You may now wish to return your calculator to its default settings (all items on the mode screen having the left item highlighted). The examples throughout this manual are worked using the default settings unless otherwise noted.

NOTES

NOTES

Appendix C

TI-85 Operations

C-1 Operations

Press ON to turn on the calculator.

Press CLEAR to clear the screen.

Press EXIT to clear any menus displayed at the bottom of the screen.

Press 2nd ▲ to make the display darker.

Press 2nd ▼ to make the display lighter.

Press 2nd + to get the RESET screen.

Then press F3 <RESET> F1 <ALL> to clear all the memories, and press F4 <YES> to clear. You must now adjust the display so you can see it. Enter a few numbers and adjust the display contrast as described above.

RAM	DELET	RESET		
ALL	MEM	DFLTS		

Mem Cleared
Defaults set

Press 2nd OFF to turn off the calculator.

Home Screen

The screen where calculations are done is called the Home Screen. You can always get to the Home Screen by pressing 2nd QUIT. Note: QUIT is above the EXIT key.

The item above the key will be referenced as if it is on the key. For example, QUIT, which is above the EXIT key, will be referenced from now on as 2nd QUIT.

2nd

This key must be used to access the calculator functions above and to the left of a key cap. A new cursor is displayed, ↑. These functions are in yellow on the calculator face.

ALPHA and 2nd ALPHA

This key is used to access the calculator functions and memory locations above and to the right of a key cap. These functions are in light blue on the calculator face.

The cursor changes to an [A] when [ALPHA] is pressed. Pressing [ALPHA] and then a letter will display an upper case letter.

The cursor changes to a [a] when [2nd] [ALPHA] is pressed. Pressing [2nd] [ALPHA] will display a lower case letter.

There is a separate memory location for the upper case letter and for the lower case letter.

[MODE]

This is found above the [MORE] key and is used to set the operating mode of the calculator. When the highlighted items are on the left, the calculator is in default mode. To choose a different mode: Use the arrow keys to choose the item you want to change to, then press the [ENTER] key. The new selection will now be highlighted. Press [2nd] [QUIT] to return to the Home Screen.

Menus

Menus are listed across the bottom of the screen. To choose a menu item press the function key immediately below the menu item. In this manual, these functions will be referred to as the key to press in a box followed by the function in angle brackets. For example, the trace function of the calculator is found on the [GRAPH] menu above the [F4] key is denoted by [F4] <TRACE>.

Error Correction

Use the [2nd] [INS] key to insert characters. ([INS] is above the [DEL] key.)

Use the [DEL] key to delete characters.

Use the arrow keys [◄] or [►] to move to the left or right.

Use the up arrow key [▲] to move up one line.

Use the down arrow key [▼] to move down one line.

[2nd] [ENTRY] is used to replay the last executed operation. ([ENTRY] is above the [ENTER] key.)

C-2 Calculating

Many calculator functions are found above key caps or in menus. For example the square root function is above the x^2 key but the absolute value is found in the [MATH] menu.

EXAMPLE 1: Calculate $|4^2 - \sqrt{34}|$.

Solution:

[2nd] [MATH] [F1] <NUM> [F5] <abs> [(] [4] [x^2] [-] [2nd] [√] [34] [)] [ENTER]. [ENTER] tells the calculator to calculate. The result displayed is 10.1690481052.

C-3 Plotting Points and Drawing Line Segments

EXAMPLE 1: (A) Plot the points (-3, 6) and (1, 5). (B) Draw the line segment having these points as the endpoints.

Solution (A):

Keystrokes	Explanation
GRAPH F3 <ZOOM> F4 <ZSTD>	We first need to set the viewing window dimensions. ZOOM F4 <Standard> automatically sets the dimensions to xMin=-10, xMax=10, xScl=1, yMin=-10, yMax=10, yScl=1 and xRes=1. This will be denoted as [-10, 10]1 by [-10, 10]1 in this manual.
EXIT EXIT	
GRAPH MORE F2 <DRAW>	Get the DRAW menu and clear the drawings from the
MORE F5 <CLDRW>	calculator screen.
2nd QUIT	
2nd CATALOG ▼ ... ▼	Get the CATALOG listing all of the functions. Use
ENTER	the down arrow, or F1 <Page↓>, until the pointer
(-) 3 , 6) ENTER	points to PtOn(. Press ENTER.
2nd ENTRY ◄ ... ◄	Enter the coordinates of the first point.
1 , 5) ENTER	Use the replay feature of the calculator to get the last entry. Use the left arrow and change the coordinates to those of the second point. Delete any unwanted characters using DEL.

Solution (B):

Keystrokes	Explanation
2nd QUIT	Return to the Home Screen.
2nd CATALOG ▼ ... ▼	Get the CATALOG listing all of the functions. Use the down arrow or F1 <PAGE↓> until the pointer
ENTER	points to Line(1. Press ENTER.
(-) 3 , 6 ,	Enter the coordinates of the first point and the second
1 , 5) ENTER	point separated by commas. Press ENTER to draw the line segment.

C-4 Storing and Evaluating an Expression, and Using Lists

EXAMPLE 1: Store the expression $\frac{-b+\sqrt{b^2-4ac}}{2a}$ and evaluate it to find one solution of $3x^2+4x-5=0$.

Solution:

Keystrokes	Explanation
GRAPH F1 <y(x)=> CLEAR ((-) ALPHA B + 2nd √ (ALPHA B x^2 – 4 ALPHA A × ALPHA C)) ÷ (2 ALPHA A)	We can store up to 99 functions or algebraic expressions in this list. Then we can recall them and evaluate or graph them. Press these keys to enter the expression.
2nd QUIT	Get back to the home screen.
3 STO▸ A ENTER 4 STO▸ B ENTER (-) 5 STO▸ C ENTER	Store 3 as A, 4 as B and -5 as C.

METHOD 1

Keystrokes	Explanation
2nd VARS F1 <ALL> F1 <PAGE↓>... F1 <PAGE↓> ▼ ENTER ENTER	Get the variables list. Press F1 <PAGE↓> and/or the arrow keys until the cursor is pointing at **y1**. Press ENTER ENTER The result is .786299647847.

METHOD 2

Keystrokes	Explanation
2nd ALPHA Y 1 ENTER	Enter y1 on the screen and press ENTER. The result is .786299647847.

EXAMPLE 2: Given $f(x) = \sqrt[3]{x^4+2x}$ and $g(x) = 4^{2x-90}$. Find $f(x) + 3\sqrt{g(x)}$ at $x = 45.2$.

Solution:

Keystrokes	Explanation
GRAPH F1 <y(x)=> CLEAR (x-VAR ^ 4 + 2 x-VAR) ^ (1 ÷ 3) ENTER CLEAR 4 ^ (2 x-VAR - 90)	Enter the functions as y1 and y2. Press CLEAR to delete the old y1 function. Note that lower case x is used in functions. Upper case and lower case letters are different. Values can be stored for each of these.
2nd QUIT	
45.2 STO▸ x-VAR ENTER	Store 45.2 as x.
2nd VARS F1 <ALL> ▼ ... ▼ ENTER + 3 2nd √ 2nd VARS F1 <ALL> ▼ ... ▼ ENTER ENTER	Form $f(x) + 3\sqrt{g(x)}$ as y1 + 3√ y2 getting y1 and y2 from the variables list. The expression will be evaluated when you press ENTER. The result is 164.97 to two decimal places.

EXAMPLE 3: Evaluate the expression $\frac{3x^2 + 5x - 8}{x^2 - 1}$ to three decimal places for x = .900, .990, .999, 1.001, 1.010, and 1.100 by: (A) evaluating the expression for each value of x individually; and (B) using a list.

Solution (A): First store the expression in the Y= list. Then store the value for x. Recall the expression and evaluate it.

Keystrokes	Explanation
[GRAPH] [F1] <y(x)=> [CLEAR]	Clear the old expression for y1 by pressing [CLEAR].
[(] [3] [x-VAR] [^] [2] [+] [5] [x-VAR] [–] [8] [)] [÷] [(] [x-VAR] [^] [2] [–] [1] [)] [2nd] [QUIT]	Store the expression as y1 and return to the home screen.
[.9] [STO▶] [x-VAR] [ENTER]	Store the value of x.
[2nd] [alpha] [Y] [1] [ENTER]	Recall the expression and evaluate.

Store the next value for x and reevaluate the expression. Repeat for the other values of x. The corresponding results are: 5.632, 5.513, 5.501, 5.499, 5.488, and 5.381, rounded to three decimal places.

Solution (B): Store the values of x in a list and evaluate the expression for each value of x simultaneously.

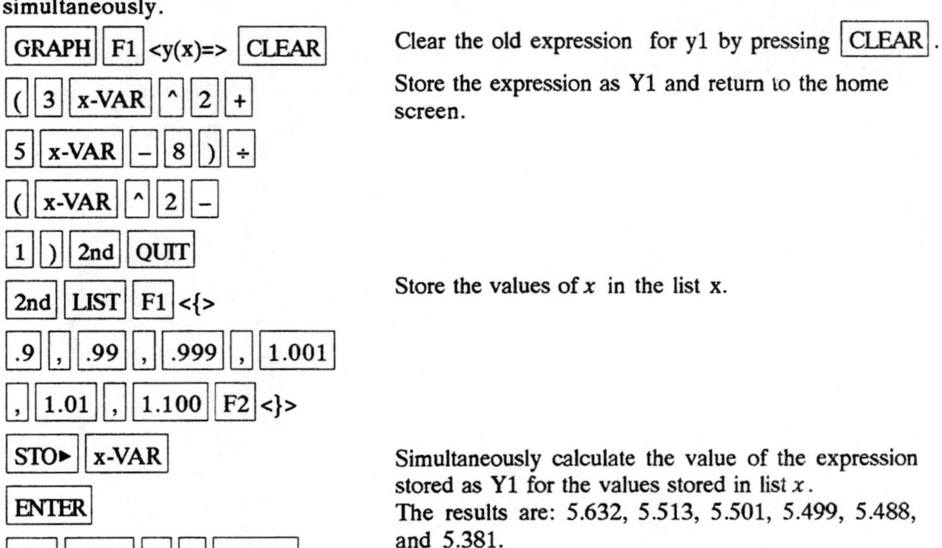

Keystrokes	Explanation
[GRAPH] [F1] <y(x)=> [CLEAR]	Clear the old expression for y1 by pressing [CLEAR].
[(] [3] [x-VAR] [^] [2] [+] [5] [x-VAR] [–] [8] [)] [÷] [(] [x-VAR] [^] [2] [–] [1] [)] [2nd] [QUIT]	Store the expression as Y1 and return to the home screen.
[2nd] [LIST] [F1] <{> [.9] [,] [.99] [,] [.999] [,] [1.001] [,] [1.01] [,] [1.100] [F2] <}> [STO▶] [x-VAR] [ENTER]	Store the values of x in the list x.
[2nd] [alpha] [Y] [1] [ENTER]	Simultaneously calculate the value of the expression stored as Y1 for the values stored in list x. The results are: 5.632, 5.513, 5.501, 5.499, 5.488, and 5.381.

C-5 Evaluating Inequalities

EXAMPLE 1: Given the inequality $x^2 - 3t < 5x + 9t^2$. Test this inequality for $x = 1.89$ and $t = -4.52$.

Solution:

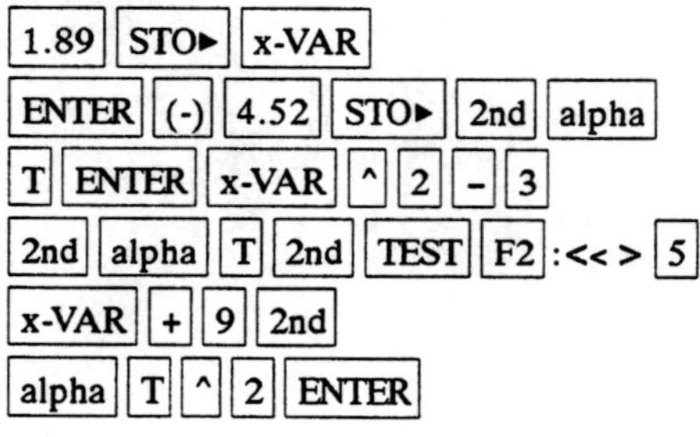

Store the value of the variables. Lower case letters are used.

Enter the expression and evaluate.

The result is 1. This means the inequality is true for these values of the variables. If the inequality was false, the result would have been 0.

C-6 Graphing

EXAMPLE 1: Graph $y = 3x^2 - 6$ and $y = 3x^2 + 1$ on the same graphing screen.

Solution:

STAT F3 <DRAW> F5 <CLDRW> — Clear drawings.

GRAPH F1 <y(x)=> CLEAR 3 x-VAR ^ 2 - 6 ENTER CLEAR 3 x-VAR ^ 2 + 1 ENTER — Store the functions. Remember to clear out the old function whenever entering a new one. Clear or deselect all others by placing the cursor at the expression for the function and pressing F5. Now this function will not graph. Note that the = sign is no longer highlighted.

EXIT — Exit the $y(x)=$ menu.

F3 <ZOOM> F4 <ZSTD> — Press F3 and then F4 to get the standard viewing window. This will automatically set the graphing screen to have $-10 < x < 10$ and $-10 < y < 10$ and graph the functions.

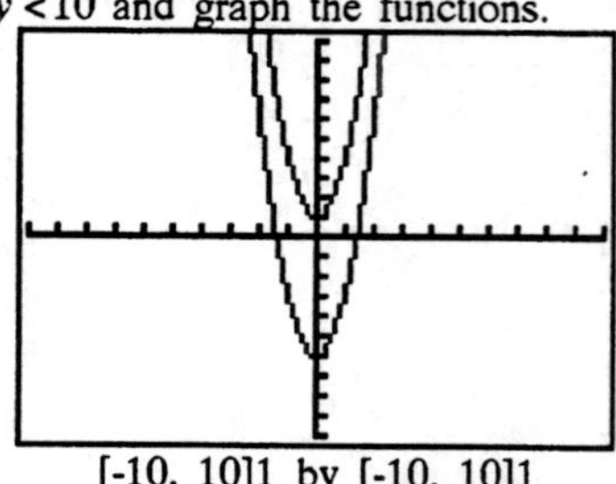

[-10, 10]1 by [-10, 10]1

CLEAR
2nd QUIT — Note that pressing CLEAR will remove the menus from the bottom of the screen.

ZSTD automatically sets the graphing screen limits to $-10 < x < 10$ and $-10 < y < 10$. These graph screen limits will be denoted in this manual by the notation [-10, 10]1 by [-10, 10]1. The graph screen limits can be set by pressing F2 <RANGE> on the GRAPH menu and entering the desired values.

F2 **<RANGE>** means to choose the RANGE option from the menu on the screen above the F2 key.

The function will <u>not</u> graph if you choose a function on the F1 <y(x)=> menu and then press F5 <SELCT>. You will notice that the highlighting disappears from the = sign. You can still evaluate this expression for stored values of the variables. To select the function, press F5 <SELCT> again.

EXAMPLE 2: Graph $f(x) = \begin{cases} .5x + 1 & x \le 2 \\ x^2 + 2 & x > 2 \end{cases}$.

Solution:

The calculator will only graph functions for values of x for which the function is defined, so we write this piecewise-defined function in a form where the pieces are defined only for the values of x that we want. Enter two functions as y1=$(.5x+1)\div(x \le 2)$ and y2= $(x^2+2)\div(x>2)$. The first function is defined with $x \le 2$ and the second function is defined when $x > 2$.

Keystrokes	
GRAPH F1 <y(x)=> CLEAR	Clear the existing function.
(.5 x-VAR + 1) ÷	Enter the first piece of the piecewise-defined function.
(x-VAR 2nd TEST	
F4 <≤> 2) ENTER	
CLEAR (x-VAR ^ 2 +	Clear the existing function. Enter the second piece of the piecewise-defined function.
2) ÷ (x-VAR 2nd	
TEST F3 <>> 2) ENTER	
EXIT EXIT F3 <ZOOM>	
F4 <ZSTD>	Graph.

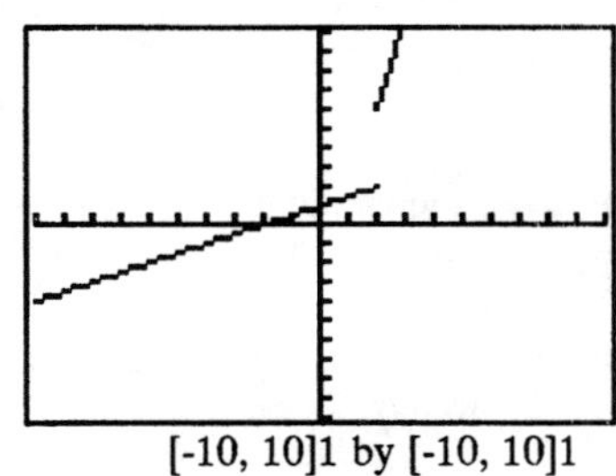

[-10, 10]1 by [-10, 10]1

C-7 TRACE, ZOOM, RANGE and SOLVER

EXAMPLE 1: Find the x intercept(s) for $x>0$ for $y = 3x^2-6$. Use three decimal place accuracy.

Solution: Store the function as 3 X ^ 2 - 6 as y1. Graph using the [-10, 10]1 by [-10, 10]1. (see Example 1 from Section D-6).

You see a new cursor near the center of the graph screen and the coordinates of that point on the graph displayed at the bottom of the screen.

Use right and left arrow keys and move the cursor along the graph. If we had the graph of two or more functions displayed, we could use the up and down arrow keys to move between the functions. The 1 in the upper right-hand corner indicates that the cursor is on the graph of y1.

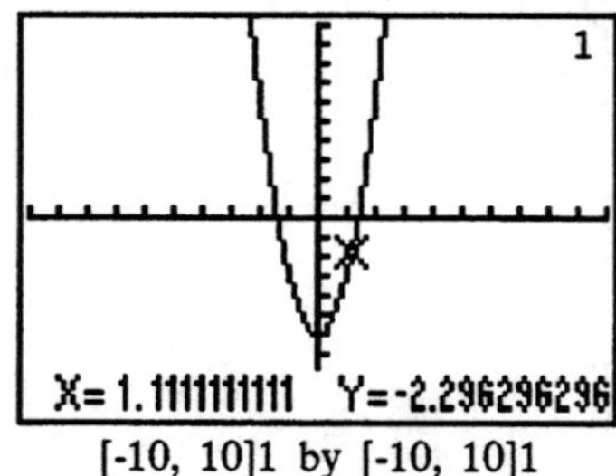

[-10, 10]1 by [-10, 10]1

There are four methods we can use to get an approximation to the x intercepts.

1. We can zoom in on the point by changing the graph screen parameters values.
2. We can zoom in on the point by settin the Zoom Factors and zoom in.
3. We can zoom in on the point by using the Zoom Box feature.
4. We can use the Solver feature .

METHOD 1 Set the RANGE

Using the trace feature of the calculator, trace the functions graphed above. We see the x intercept is near 1.5. So we shall set the RANGE closer to that value.

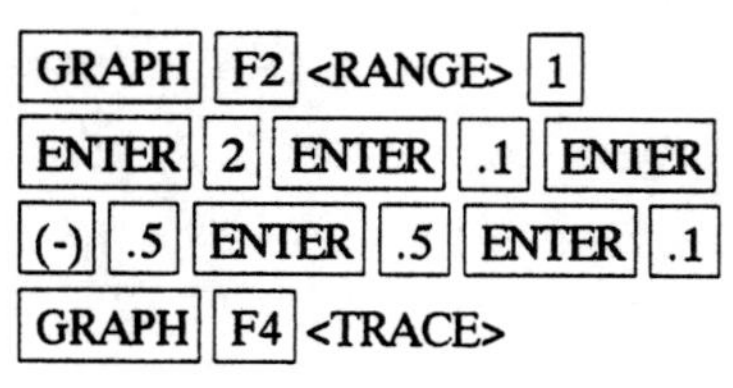

Set the RANGE to [1, 2].1 by [-.5, .5].1.

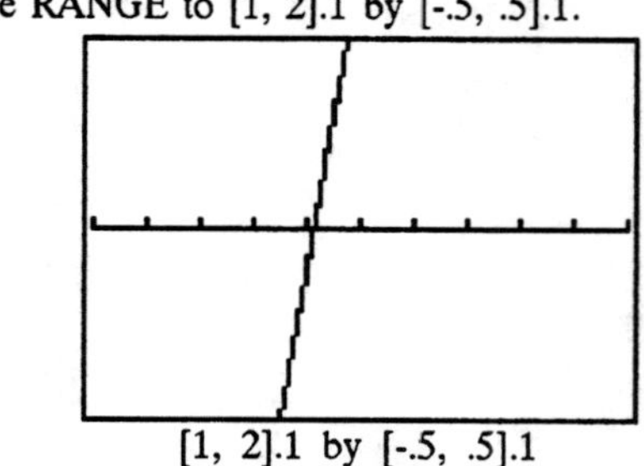

[1, 2].1 by [-.5, .5].1

Press EXIT to restore the menu at the bottom of the screen.

Use TRACE again to get a better approximation. The trace cursor always starts at the middle x value of the RANGE. It may not be on the display. Watch the values for x at the bottom of the screen.

We see that the x intercept is between 1.41 and 1.42. Set the RANGE to [1.41, 1.42].01 by [-.01, .01].001 and graph again. We now find the x intercept to 1.414 accurate to three decimal places.

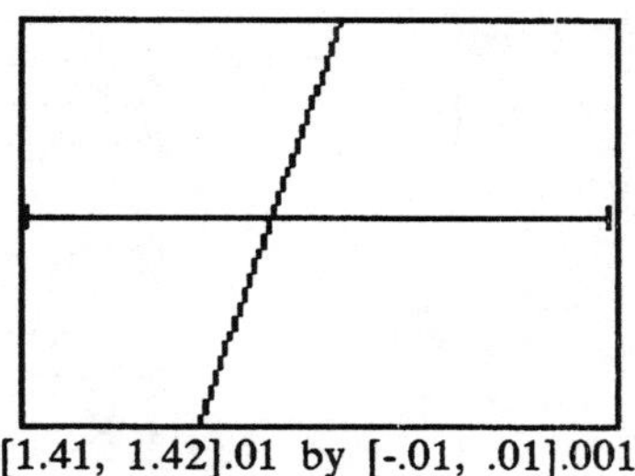

[1.41, 1.42].01 by [-.01, .01].001

METHOD 2 Zoom In

Keystrokes	
GRAPH F3 <ZOOM> F4 <ZSTD>	Get the [-10, 10]1 by [-10, 10]1 display as shown at the beginning of this section.
MORE MORE F1 <ZFACT> 5 ENTER 5	Set the Zoom Factors to 5 and 5. The graph will be 5 times as large in both the x and y directions.
ENTER F3 <ZOOM> F2 <ZIN> ◄ or ► or ▲ or ▼	Use the ◄ or ► arrow keys to get the cursor to where you want the center of the new display, say at (1.42, 0).
EXIT	Press EXIT to restore the menu at the bottom of the screen.
ENTER	The new display will be centered where you placed the cursor.

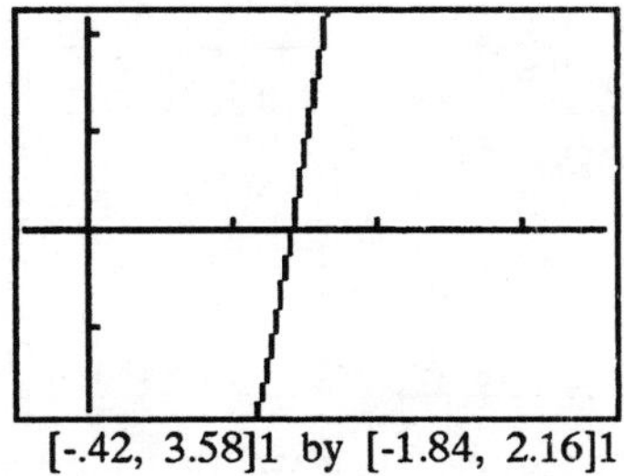

[-.42, 3.58]1 by [-1.84, 2.16]1

Now trace again to get a new approximation to the x intercept.
Repeat the zooming in process as many times as necessary to get accuracy to three decimal places. The x intercept is 1.414 .

METHOD 3 Zoom Box

GRAPH F3 <ZOOM> — Get the [-10, 10]1 by [-10, 10]1 display as shown at the beginning of this section.

F4 <ZSTD>

F1 <BOX> — Get the Zoom Box option.

◄ or ► or ▲ or ▼
ENTER

◄ or ► or ▲ or ▼
ENTER

Use the arrow keys repeatedly to get the cursor close to but a little above and to the left of the x intercept, say at X = 1.1111111111, Y = .9677419355. Press ENTER to place the upper corner of the Zoom Box at this point. Now move the cursor down and to the right of the x intercept, say at X = 2.0634920635 Y = -.6451612903. Press ENTER again. The graph will be drawn again with the box as the outside of the new display.

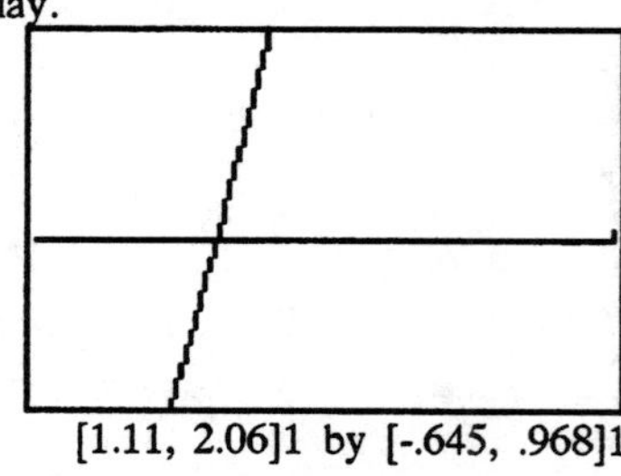

[1.11, 2.06]1 by [-.645, .968]1

Use Trace again to get another estimate of the x intercept. Use Zoom Box a second time to get a better approximation of the x intercept. The x intercept is 1.414 .

METHOD 4 Solver

2nd SOLVER F1 <y1> alpha = 0 ENTER — Get the Solver feature.

1.5 F5 <SOLVE> — Enter an estimate of the x intercept and solve. The x intercept is 1.414.

C-8 Choosing the RANGE Variables Values

There is no one right way to determine the RANGE variables values that should be used for the graph of a function. Three methods to try are:

1. Graph using the standard screen and then zoom out.
2. Analyze the features of the function and then choose a RANGE.
3. Evaluate the function at several values.

EXAMPLE 1 Graph $y = .03x^3 + .02x^2 - 5x + 10$.

Solution:

METHOD 1 Graph using the standard screen and then zoom out

Enter the function as .03 X ^ 3 + .02 X ^ 2 - 5 X + 10.

Press EXIT F3 <ZOOM> F4 <ZSTD> to get the [-10, 10]1 by [-10, 10]1 display.

Press F3 <ZOOM> MORE MORE F1 <ZFACT> and set the Zoom Factors to xFact = 4 and yFact=4. Press F3 <ZOOM> F3 <ZOUT>.

Move the cursor so it is where you want the center of the new screen to be, and press ENTER. The result is shown on the next page. The screen has been zoomed out to give us a better view of the graph. Note the double lines as the axes. This is because the Xscl and Yscl values are too close together. They do not change when zooming in.

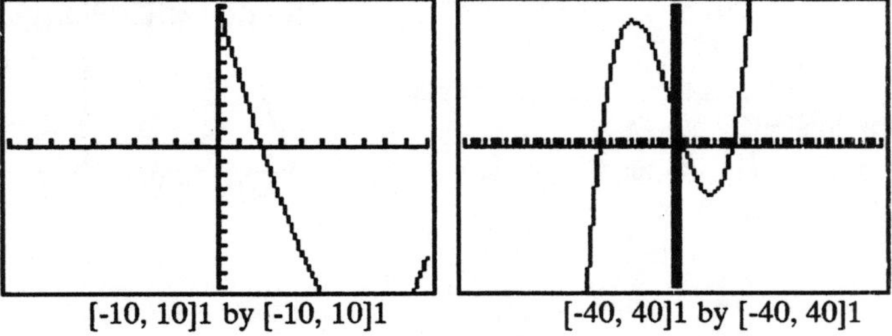

[-10, 10]1 by [-10, 10]1 [-40, 40]1 by [-40, 40]1

METHOD 2 Analyze the function and then choose the window variable values

We will analyze the term with highest power on x since this is the dominant term of the function.

We see that the leading coefficient is .03. This means that the vertical change in the graph will be approximately .03 unit for every time x^3 increases 1 unit. An equation that represents this is $.03x^3 = y$ or $x^3 = \frac{1}{.03}y$ or $x^3 \approx 30y$. So if y increases by 1 unit, x will increase by about $\sqrt[3]{30}$ or about 3 units. This indicates that we want the x axis three times as wide as the y axis. Hence, a good first choice for the horizontal axis is -30 to 30, and a good first choice for the vertical axis is -10 to 10. Set Xmin = -30, Xmax = 30 and Xscl = 3. We choose 3 as the Xscl because this will put 20 tick marks on the axis.

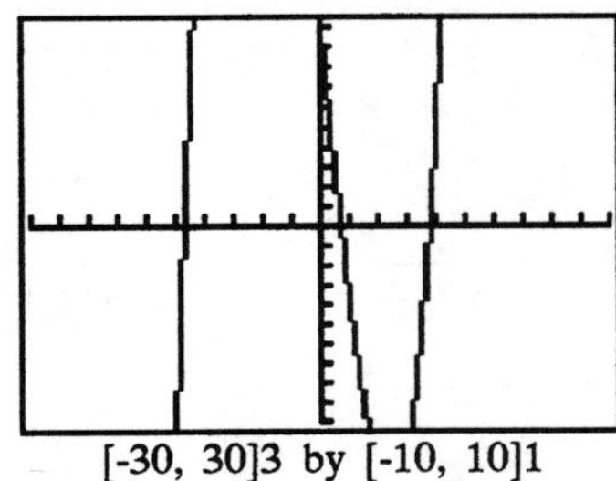

[-30, 30]3 by [-10, 10]1

The values for the vertical variables needs some adjusting since the turning points of the graph are not shown on the screen. Window variable values of [-30, 30]1 by [-40, 40]5 would show the turning points of the graph.

METHOD 3 Evaluate the function at several values of x and then choose a RANGE. The choice of x values is arbitrary. Store the value of x in memory location x, then evaluate the function. Repeat for all values of x.

x	$f(x)$
-25	-321.25
-10	32.00
0	10.00
10	-8.00
25	366.25

Choose the Xscl and Yscl so there are about 20 marks on each axis (see METHOD 1 above).

So a good first choice for the RANGE could be [-25, 25]2 by [-300, 300]30.

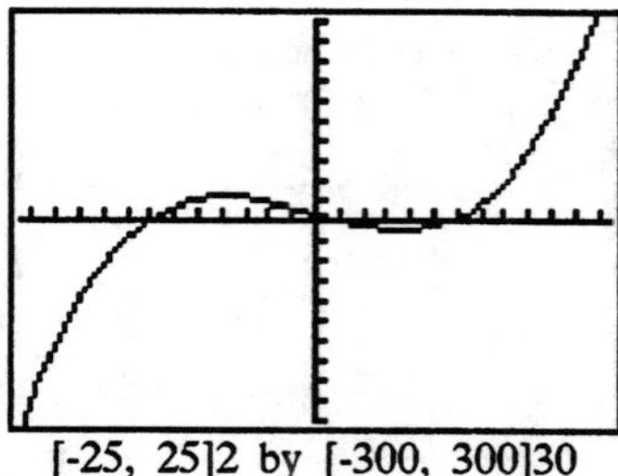

[-25, 25]2 by [-300, 300]30

C-9 Equations in One Variable

An equation in one variable can be solved graphically in two ways:

1. Algebraically move all terms to one side of the equation so that you have an expression equal to zero. Create a function $f(x)$ = (the expression). Graph the function.

   There are three ways to solve an equation in the form $f(x) = 0$ or to find the x intercepts of the graph of $f(x)$:

   a. Trace and zoom

   b. Use the solver

   c. Use the root finding feature

   The set of x intercepts is the solution to the equation.

2. Create two functions: $f(x)$ = (the left side of the equation) and $g(x)$ = (the right side of the equation). Graph both functions. Trace and zoom, or use the solver, to find the x coordinate of the points of intersection of the two graphs. The set of x coordinates is the solution to the equation.

EXAMPLE: Solve $\sqrt[3]{x+2} = 4x^2 - 3\sqrt{x+4}$. Use two decimal places.

Solution:

METHOD 1 Write the equation as $\sqrt[3]{x+2} - (4x^2 - 3\sqrt{x+4}) = 0$. Create the function

$f(x) = \sqrt[3]{x+2} - (4x^2 - 3\sqrt{x+4})$. Enter this function into the calculator and graph.

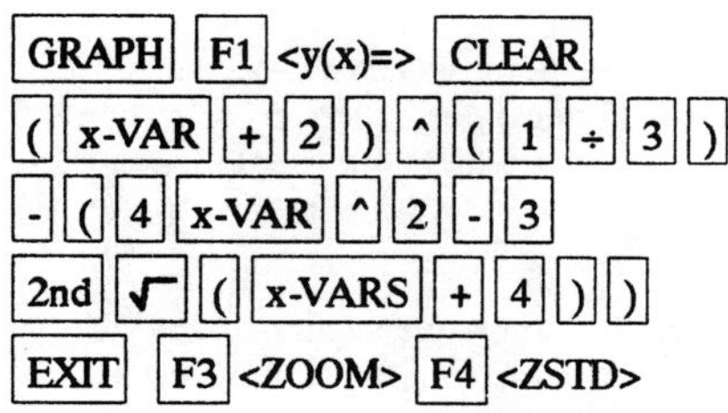

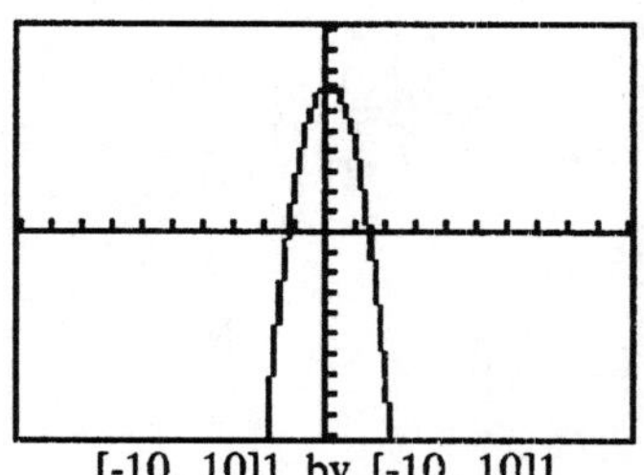

[-10, 10]1 by [-10, 10]1

We see the function crosses the x axis in two places.

a. Trace and zoom. This method is described in section C-8.

b. Solver.

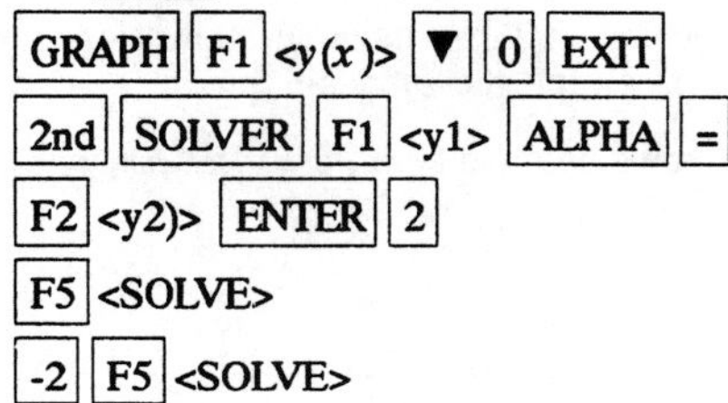

Get the y(x)= menu and enter 0 as one of the functions. This is the equation of the x axis. Now set Now set y1 equal to y2. Enter a number near the x intercept you are finding. We used $x = 2$. Solve.
Enter another value of x near the other x intercept. We used -2. Solve.
The two intercepts are $x = -1.22$ and $x = 1.46$.

c. Root.

GRAPH MORE F1 <MATH> F3 <ROOT>
◀ or ▶ ENTER

Get the MATH menu from GRAPH menu at the bottom of the screen after you press GRAPH. Choose ROOT and move the cursor close to the x intercept. Press ENTER to get the root.

METHOD 2 Create the two functions $f(x) = \sqrt[3]{x + 2}$ and $g(x) = 4x^2 - 3\sqrt{x+4}$. Graph these and find the x coordinates of the intersection points. This can be done using trace and zoom or using the solver. The solver was used in the description below.

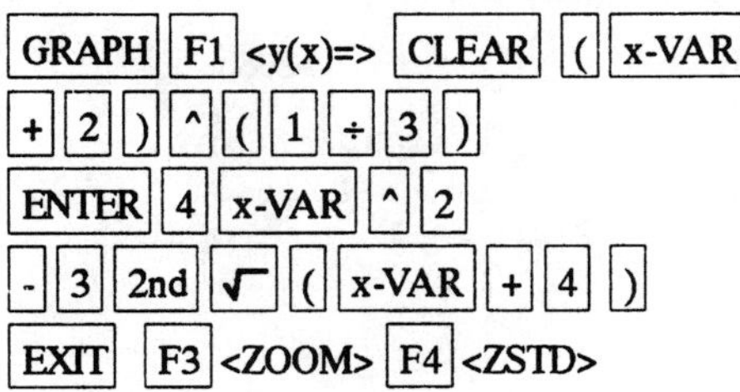

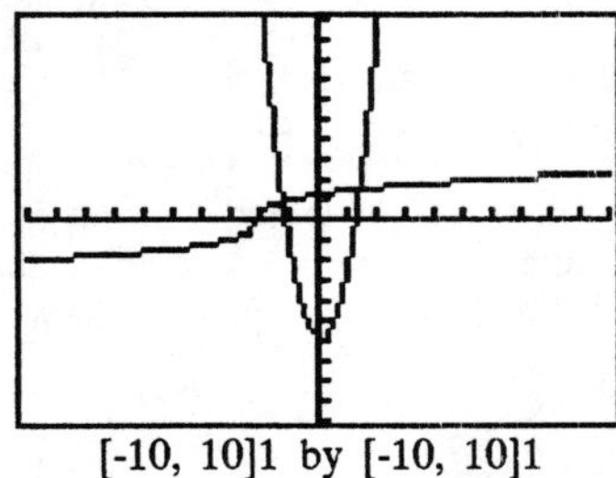

[-10, 10]1 by [-10, 10]1

[2nd] [SOLVER] [F1] <y1>

[ALPHA] [=] [F2] <y2)> [ENTER] [-2]

[F5] <SOLVE>

You'll need to enter a value of x near the x coordinate of one intersection point so the calculator knows which intersection point you are solving for. We chose 2. The result is $x = 1.46$, $y = 1.51$.

[2] [F5] <SOLVE>

Repeat for the other intersection point. We used -2 as the approximate value of this x coordinate. The result is $x = -1.22$, $y = .92$.

So the solution to this equation is $x = -1.22$ and $x = 1.46$.

C-10 Inequalities in One Variable

An inequality in one variable can be solved graphically in two ways:

1. Algebraically move all terms to one side of the equation so that you have an expression equal to zero. Create a function $f(x)$ = (the expression). Graph the function.

   There are three ways (as described in Section C-9) to solve an equation in the form $f(x) = 0$ or to find the x intercepts of the graph of $f(x)$:

   a. Trace and zoom

   b. Use the solver

   c. Use the root finding feature

   The set of x intercepts are the cutoff values for the solution. Now determine the intervals over which the inequality is true and write the solution.

2. Create two functions: $f(x)$ = (the left side of the equation) and $g(x)$ = (the right side of the equation). Graph both functions. Find the x coordinate of the points of intersection of the two graphs. The set of x coordinates are the cutoff values for the solution. Now, determine the intervals over which the inequality is true and write the solution.

EXAMPLE: Solve $\sqrt[3]{x+2} < 4x^2 - 3\sqrt{x+4}$. Use two decimal places.

Solution: METHOD 1 Write the inequality as: $\sqrt[3]{x+2} - (4x^2 - 3\sqrt{x+4}) < 0$.

Change the sign to an equal sign: $\sqrt[3]{x+2} - (4x^2 - 3\sqrt{x+4}) = 0$. Create the function $f(x) = \sqrt[3]{x+2} - (4x^2 - 3\sqrt{x+4})$, graph, and find the x intercepts. This was done above in Appendix Section C-9 Method 1 of this manual. The cutoff values are $x = -1.22$ and $x = 1.46$. Observing the graph, we see it is below the x axis for $x < -1.22$ and $x > 1.46$. Hence the solution to this inequality can be written as
$(-\infty, -1.22) \cup (1.46, \infty)$.

METHOD 2 Create the two functions $f(x) = \sqrt[3]{x+2}$ and $g(x) = 4x^2 - 3\sqrt{x+4}$. Graph these and find the x values of the intersection points. This was done above in Appendix Section C-9 Method 2 of this manual. The intersection points are at (-1.22, .92) and (1.46, 1.51). Inspecting the graph we see that $f(x) < g(x)$ for $x < -1.22$ and $x > 1.46$. Hence the solution to this inequality can be written as $(-\infty, -1.22) \cup (1.46, \infty)$.

C-11 Permutations, Combinations, and Random Numbers

EXAMPLE 1: Find: (A) $P_{19,13}$ and (B) $C_{19,13}$.

Solution (A): This is a direct calculation.

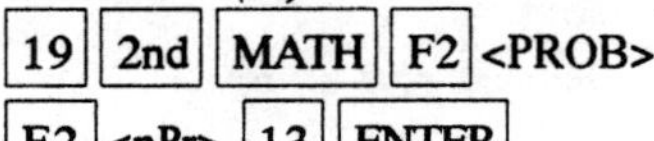

Keystrokes	
[19] [2nd] [MATH] [F2] <PROB> [F2] <nPr> [13] [ENTER]	Enter the numbers and get the built-in function from the MATH menu. The result is displayed in scientific notation as 1.68951528346E14.

Solution (B): This is a direct calculation.

Keystrokes	
[19] [2nd] [MATH] [F2] <PROB> [F3] <nCr> [13] [ENTER]	Enter the numbers and get the built-in function from the MATH menu. The result is 27132.
[EXIT] [EXIT]	To return to the home screen.

EXAMPLE 2: Generate ten unique random numbers at least 0 and less than 50.

Solution:

Keystrokes	
[5] [STO▸] [2nd] [MATH] [F2] <PROB> [F4] <rand> [ENTER]	We need to store a "seed" number in the storage location RAND. This tells the random number generator where to start.
[EXIT] [F1] <NUM> [F2] <IPart> [(] [50] [x] [EXIT] [F2] <PROB> [F4] <rand> [)] [ENTER] [ENTER]	RAND generates a number greater than 0 and less than 1. If we multiply by 50 and take the integer part, the result is a number greater than or equal to 0 and less than 50.
[ENTER] [ENTER] [ENTER] [ENTER] [ENTER] [ENTER] [ENTER] [ENTER]	The 10 random numbers are: 36 13 6 17 21 ~~21~~ 20 5 31 16 48. Not that repetitions can occur but 21 cannot be used twice if want unique random numbers.

C-12 Matrices

Given the matrices $A = \begin{bmatrix} 1 & -4 & 6 \\ 2 & -1 & 0 \\ 5 & -3 & 6 \end{bmatrix}$ and $B = \begin{bmatrix} 4 & 15 \\ -2 & 6 \\ 2 & 5 \end{bmatrix}$.

EXAMPLE 1: Find -2.5AB.

Solution:

Keystrokes	
[2nd] [MATRX] [F2] <EDIT> [A] [ENTER]	Choose matrix A and enter the dimensions.
[3] [ENTER] [3] [ENTER]	
[1] [ENTER] [(-)] [4] [ENTER] [6] [ENTER]	Enter the matrix elements row by row.
[2] [ENTER] [(-)] [1] [ENTER] [0] [ENTER]	
[5] [ENTER] [(-)] [3] [ENTER] [6] [EXIT]	
[2nd] [MATRX]	Choose matrix B and enter the dimensions.
[F2] <EDIT> [B] [ENTER]	
[3] [ENTER] [2] [ENTER]	
[4] [ENTER] [15] [ENTER]	
[(-)] [2] [ENTER] [6] [ENTER]	Enter the matrix elements.
[2] [ENTER] [5]	
[EXIT]	Return to the Home Screen. Enter the matrix operations.
[(-)] [2.5] [2nd] [MATRX]	
[F1] <NAMES> [F1] <A> [×] [F2] <B>	The result is $\begin{bmatrix} -60 & -52.5 \\ -25 & -60 \\ -95 & -217.5 \end{bmatrix}$.
[ENTER]	

EXAMPLE 2: Find A^{-1}.

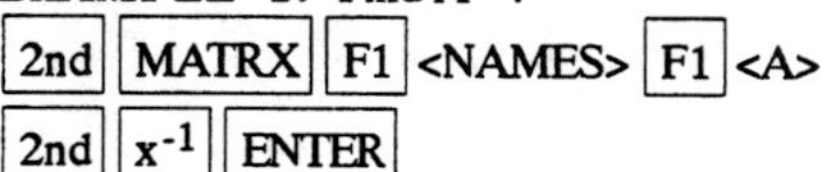

[2nd] [MATRX] [F1] <NAMES> [F1] <A>

[2nd] [x^{-1}] [ENTER]

The operations are done from the Home Screen. The result rounded to three decimal places is:

$$A = \begin{bmatrix} -.167 & .167 & .167 \\ -.333 & -.667 & .333 \\ -.028 & -.472 & .194 \end{bmatrix}.$$

Some numbers may look like decimals. However upon examining the matrix using the arrow keys you will find they are decimals that are very close to zero. This situation can be eliminated by setting the number of decimal places to a number, say 3 (see Section C-3 MODES and/or C-17 of this manual).

EXAMPLE 3: Solve the system of equations using the inverse of the coefficient matrix. Round answers to two decimal places.

$$\begin{aligned} x - 4y + 6z &= 2.5 \\ 2x - y \quad\quad &= 3.2 \\ 5x - 3y + 6z &= -3.4 \end{aligned}$$

Solution: The inverse of the coefficient matrix was found above in Section C-12 Example 2. Enter the constants as matrix B and find $A^{-1}B$.

[2nd] [MATRX]
[F2] <EDIT> [2] :<B> [ENTER]
[3] [ENTER] [1] [ENTER] [2.5] [ENTER]
[3.2] [ENTER] [(-)] [3.4] [EXIT]

Enter the matrix dimensions and enter the matrix elements.

[EXIT] [2nd] [MATRX] [F1] <NAMES>
[F1] <A> [2nd] [x^{-1}] [×] [F2] <B> [ENTER]

Return to the Home Screen. Perform the matrix operations.

The result is $\begin{bmatrix} -.45 \\ -4.1 \\ -2.24 \end{bmatrix}$. The solution to the system of equations is $x = -.45$, $y = -4.1$, $z = -2.24$.

[EXIT] [EXIT]

Return to the home screen.

EXAMPLE 4: Find the solution to the system of Example 3 above using Gauss-Jordan reduction. In other words, find the reduced form of the matrix $\begin{bmatrix} 1 & -4 & 6 & 2.5 \\ 2 & -1 & 0 & 3.2 \\ 5 & -3 & 6 & -3.4 \end{bmatrix}$.

Solution:

METHOD 1 Use row operations on the matrix to get the reduced form.

[2nd] [MATRX] [F2] <EDIT>
[A] [ENTER] [3] [ENTER] [4] [ENTER]

Enter the matrix dimensions.

[1] [ENTER] [(-)] [4] [ENTER] [6]
[ENTER] [2.5] [ENTER] [2] [ENTER]
[(-)] [1] [ENTER] [0] [ENTER] [3.2]
[ENTER] [5] [ENTER] [(-)] [3] [ENTER]
[6] [ENTER] [(-)] [3.4] [EXIT] [EXIT]

Enter the matrix elements. Notice that if you are continuing from Example 1 - 3 of this section, you need only enter the elements in column 4. Use the down arrow to move from one row to the next.

[2nd] [MATRX] [F4] <OPS>
[MORE] [F5] <mRAdd>
[(-)] [2] [,] [EXIT] [F1] <NAMES>
[F1] <A> [,] [1] [,] [2] [)] [ENTER]

We wish to multiply by -2, matrix A, row 1 and add it to row 2. The result will be put in row 2.

```
mRadd(-2,A,1,2)
```

The result is $\begin{bmatrix} 1 & -4 & 6 & 2.5 \\ 0 & 7 & -12 & -1.8 \\ 5 & -3 & 6 & -3.4 \end{bmatrix}$.

[STO▸] [F1] <A>

Store the result in matrix A.

[ENTER] [2nd] [MATRX]
[F4] <OPS> [MORE] [F5] <mRAdd>
[(-)] [5] [,] [EXIT] [F1] <NAMES>
[F1] <A> [,] [1] [,] [3] [)] [ENTER]

Now multiply row 1 by -5 and add it to row 3 putting the result in row 3.

```
mRadd(-5,A,1,3)
```

The result is $\begin{bmatrix} 1 & -4 & 6 & 2.5 \\ 0 & 7 & -12 & -1.8 \\ 0 & 17 & -24 & -15.9 \end{bmatrix}$.

Keystrokes	Explanation
[STO▸] [A] [ENTER]	Store the result in matrix A.
[2nd] [MATRX] [F4] <OPS> [MORE] [F4] <multR> [1] [÷] [7] [,] [ALPHA] [A] [,] [2] [)] [ENTER] [STO▸] [A]	Multiply row 2 by 1/7 . The result is $\begin{bmatrix} 1 & -4 & 6 & 2.5 \\ 0 & 1 & -1.71 & -.26 \\ 0 & 17 & -24 & -15.9 \end{bmatrix}$ rounded to two decimal places. Store the result in matrix A.
	Continue in this manner until you get $\begin{bmatrix} 1 & 0 & 0 & -.45 \\ 0 & 1 & 0 & -4.1 \\ 0 & 0 & 1 & -2.24 \end{bmatrix}$. The solution to the system of equations is $x = -.45$, $y = -4.1$, $z = -2.24$.

> Note: To swap rows of a matrix use [F4] <OPS> [MORE] [F2] <rSwap> from the [MATRX] menu. The command rSwap(A,2,3) will swap rows 2 and 3 in matrix A.

<u>METHOD 2</u> Use the built-in capabilities of the calculator to get the reduced form.

Keystrokes	Explanation
[2nd] [MODE] [▼] [▶] [▶] [▶] [ENTER]	Set the mode to two decimal places.
[2nd] [MATRX] [F2] <EDIT>	Enter the matrix dimensions.
[A] [ENTER] [3] [ENTER] [4] [ENTER] [1] [ENTER] [(-)] [4] [ENTER] [6] [ENTER] [2.5] [ENTER] [2] [ENTER] [(-)] [1] [ENTER] [0] [ENTER] [3.2] [ENTER] [5] [ENTER] [(-)] [3] [ENTER] [6] [ENTER] [(-)] [3.4] [EXIT]	Enter the matrix elements. Notice that if you are continuing from Example 1 - 3 of this section, you need only enter the elements in column 4. Use the down arrow to move from one row to the next.
[2nd] [MATRX] [F4] <OPS> [F5] <rref> [EXIT] [F1] <NAMES> [F1] <A> [ENTER]	We wish to get the reduced form of this matrix. Use the matrix operation rref to get this on the calculator.
	The result is $\begin{bmatrix} 1.00 & 0.00 & 0.00 & -.45 \\ 0.00 & 1.00 & 0.00 & -4.10 \\ 0.00 & 0.00 & 1.00 & -2.24 \end{bmatrix}$. Use the right and left arrow keys to view the entire matrix.
	Hence, the solution to the system of equations, rounded to two decimal places, is $x = -.45$, $y = -4.10$, $z = -2.24$.

C-13 Logarithmic, Exponential, and Hyperbolic Functions

EXAMPLE 1: Graph $f(x) = 10^{3.2x}$ for $-.5 < x < .5$.

Solution: Set the RANGE to [-.5,.5].1 by [-10,10]1.

GRAPH F1 <y(x)=>

CLEAR 10 ^ (3.2 x-VAR) EXIT

F2 <RANGE> (-) .5 ENTER

.5 ENTER .1 ENTER (-) 10 ENTER

10 ENTER

1 F5 <GRAPH>

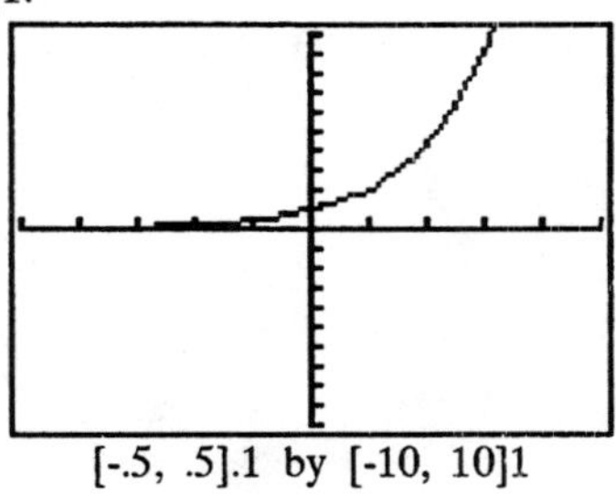

[-.5, .5].1 by [-10, 10]1

EXAMPLE 2: Graph $y = \frac{e^{x} + e^{-x}}{2}$.

Solution: This could be entered using (2nd e^x x-VAR + 2nd e^x (-) x-VAR) ÷ 2 . However the built-in hyperbolic cosine function can also be used.

GRAPH F1 <y(x)=> CLEAR

2nd MATH F4 <HYP> F2 <cosh>

x-VAR EXIT F3 <ZOOM>

F4 <ZSTD>

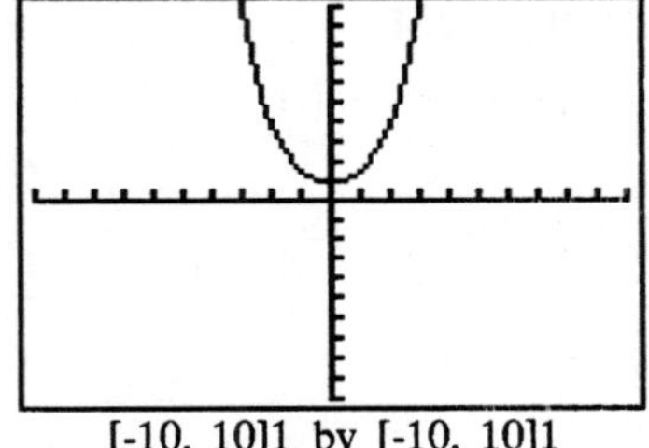

[-10, 10]1 by [-10, 10]1

EXAMPLE 3: Solve $\ln(x^2 + x) = 3x + 2$

Solution:

GRAPH F1 <y(x)=> CLEAR — Enter the function $f(x) = \ln(x + x^2)$ as Y1 in the function list.

LN (x-VAR ^ 2

+ x-VAR) ENTER — Enter the function $g(x) = 3x + 2$ as Y2 in the function list.

CLEAR 3 x-VAR + 2

ENTER

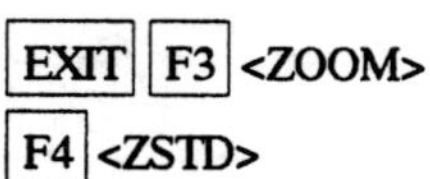

EXIT F3 <ZOOM>
F4 <ZSTD>

Graph them on the same graphing screen. Zoom and trace to find the x value of the point of intersection. This will be the solution to the equation.

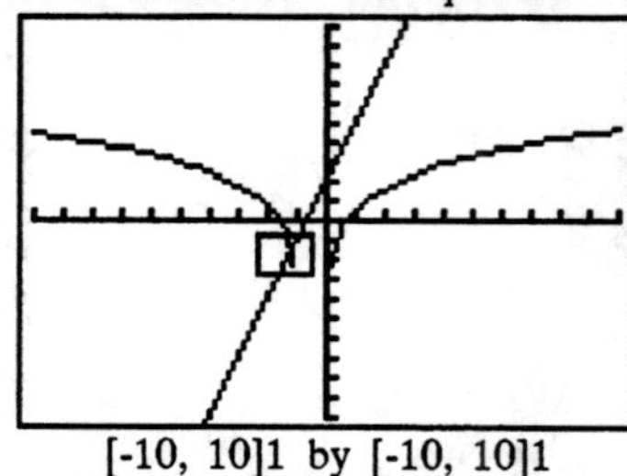

[-10, 10]1 by [-10, 10]1

The solution is -1.18, rounded to two decimal places.

C-14 Trigonometric Functions

EXAMPLE 1: Evaluate $f(x) = \cos x + \csc x$ at (A) 53°45'18" and (B) 78.52 radians.

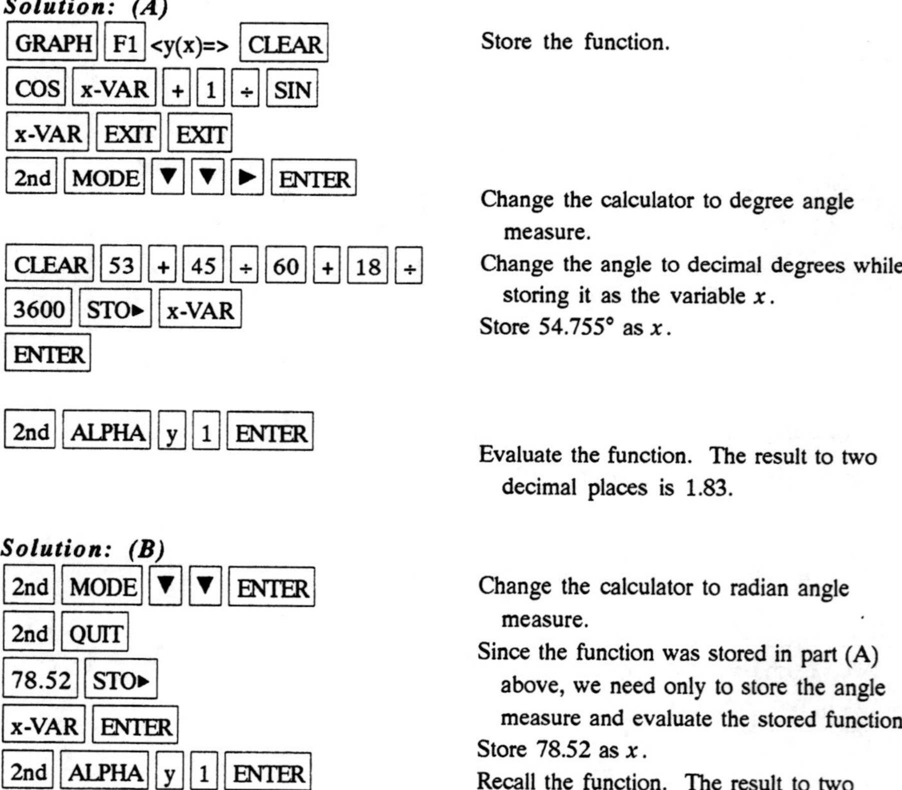

Solution: (A)

Keystrokes	Explanation
GRAPH F1 <y(x)=> CLEAR COS x-VAR + 1 ÷ SIN x-VAR EXIT EXIT	Store the function.
2nd MODE ▼ ▼ ▶ ENTER	Change the calculator to degree angle measure.
CLEAR 53 + 45 ÷ 60 + 18 ÷ 3600 STO▶ x-VAR ENTER	Change the angle to decimal degrees while storing it as the variable x. Store 54.755° as x.
2nd ALPHA y 1 ENTER	Evaluate the function. The result to two decimal places is 1.83.

Solution: (B)

Keystrokes	Explanation
2nd MODE ▼ ▼ ENTER	Change the calculator to radian angle measure.
2nd QUIT	Since the function was stored in part (A) above, we need only to store the angle measure and evaluate the stored function.
78.52 STO▶ x-VAR ENTER	Store 78.52 as x.
2nd ALPHA y 1 ENTER	Recall the function. The result to two decimal places is 49.47.

EXAMPLE 2: Graph $f(x) = \cos x + \csc x$.

Solution: The function was entered in the calculator in Example 1 above.

Keystrokes	Comments
GRAPH MORE F3 <FORMT>	Change the graph mode to Dot. This will eliminate the "vertical lines."
▼ ▼ ► ENTER	
F5 <GRAPH> F3 <ZOOM>	This option automatically sets the graph RANGE to [-8.24,8.24]1.57 by [-4,4]1.
MORE F3 <ZTRIG>	

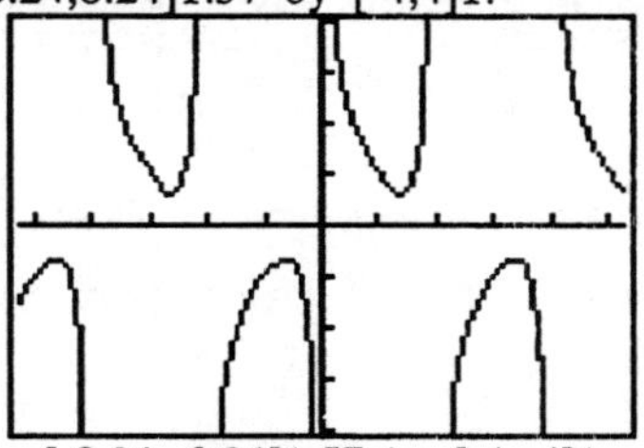

[-8.24, 8.24]1.57 by [-4, 4]1

C-15 Numeric Differentiation and Integration, and Tangent Lines

EXAMPLE 1: (A) Find the equation of the tangent line to $f(x) = \sqrt{x^5 - \sqrt[3]{x}}$ at $x = 3.54$. Use two decimal place accuracy. (B) Sketch the function and its tangent line using [2, 5]1 by [10, 60]5.

Solution (A): The function and it's derivative can be evaluated at this value of x .

Keystrokes	Comments
GRAPH F1 <y(x)=> CLEAR 2nd	Enter the function and store it as y1.
√ (x-VAR ^ 5 - x-VAR ^	
(1 ÷ 3))	
2nd QUIT 3.54 STO►	Get the Home Screen and store the value of x.
x-VAR ENTER	
2nd VARS F1 <ALL>	Use the arrow keys to highlight the function.
▼ ... ▼ ENTER ENTER	Press ENTER to evaluate the function. $f(3.54) = 23.55$.
2nd CALC F3 <der1> 2nd alpha	Get the numeric derivative built-in function: `der1(y1,x,3.54)`
y 1 , x-VAR , 3.54)	This will evaluate the derivative with respect to x of the function stored in y1 at $x = 3.54$. The result is 16.67. (The default step size used in x increments is .001.)
ENTER	

The equation of the tangent line is $y - 23.55 = 16.67(x - 3.54)$.

Solution (B): Method 1 Enter the function as Y1 in the calculator and the equation of the tangent line as Y2 in the calculator and graph.

Method 2 Enter the function as y1 in the calculator. Set the viewing rectangle to [2, 5]1 by [10, 60]5 and graph the function.

Use the tangent drawing feature of the calculator by entering [GRAPH] [MORE] [F2] <DRAW> [MORE] [MORE] [F1] <TanLn> [2nd] [ALPHA] [Y] [1] [,] [3.54] [)] [ENTER].

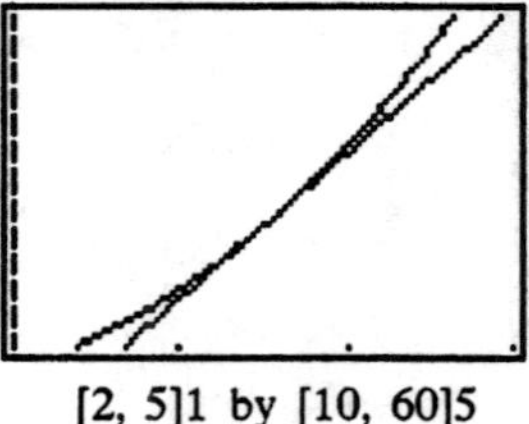

[2, 5]1 by [10, 60]5

(It may take a little while to graph.) Press [CLEAR] to remove the menu at the bottom of the screen. Press [EXIT] to display it again. The function and its tangent will be graphed as shown to the right.

EXAMPLE 2: Is the function $f(x)$ defined in Example 1 concave upwards at $x = 3.54$?

Solution:

[2nd] [CALC] [F4] <der2> [2nd] [alpha] [y] [1] [,] [x-VAR] [,] [3.54] [)] [ENTER]

Enter the function and store it as y1 as shown in Example 1 of this section.
Use the second derivative function.

```
der2(y1,x,3.54)
```

The result is positive. Hence the function is concave upwards at $x = 3.54$.

EXAMPLE 3: Find $\int_{1.32}^{2.58} \sqrt{x^5 - \sqrt[3]{x}}\, dx$. Use two decimal place accuracy.

Solution:

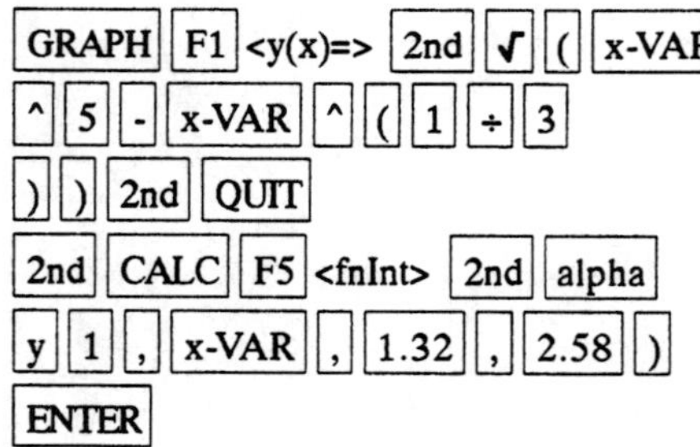

Enter the function and store it.

Get the numeric integration function built into the calculator function:

```
fnInt(y1,x,1.32,2.58)
```

This will evaluate the definite integral. The result is 6.95. The tolerance in calculating the definite integral is .00001. This can be changed by pressing [2nd] [TOLER] and changing the setting.

C-16 Statistics

EXAMPLE 1: Given the following data, find: (A) the mean, (B) the sample standard deviation, (C) an histogram with first class beginning at 10 and width 4 (use two decimal places), and (D) repeat Parts A and B after deleting the data value 89 from the list. How did eliminating this outlier effect the mean and standard deviation?

23	56	73	12	89	52	45
38	18	21	15	37	42	51
19	40	57	38	54	60	39

Solution (A) and (B):

Keystrokes	Explanation
[STAT] [F2] <EDIT>	Get the statistics menu.
[ENTER] [ENTER] [F5] <CLRxy>	Clear the statistical registers.
[23] [ENTER] [ENTER] [38] [ENTER] [ENTER] [19] [ENTER] [ENTER] Etc.	Enter the data in the x1 position. The y1 position is used for frequency if there is more than one piece of data with the same number. Enter all data. There are 21 values. If you make a mistake, use the arrow keys to highlight the value and type it in again. To remove a data value, use the arrow keys to get the x data value and press [F1] <DELi>.
[EXIT]	Exit the data entry screen.
[F1] <CALC> [ENTER] [ENTER] [F1] <1-VAR>	The mean is 41.86 and the sample standard deviation is 19.82.

Solution (C):

Keystrokes	Explanation
[EXIT] [EXIT] [GRAPH] [F1] <y(x)=> [F5] <SELCT>	This will deselect the function the cursor is on. The = sign will no longer be highlighted. Deselect all functions stored or delete them using [CLEAR].
[EXIT] [F2] <RANGE>	
[0] [ENTER] [100] [ENTER] [4] [ENTER] [(-)] [5] [ENTER] [5] [ENTER] [1] [EXIT]	The smallest value is 12 so let xMin=0. The greatest value is 89. Let us set xMax=100. Let's use xScl=4. This sets the histogram class width. Let yMin=-5 to get some room below the histogram. We do not know the frequencies yet so let yMax=5 for a first try. Let yScl=1.
[STAT] [F3] <DRAW> [F5] <CLDRW>	Clear all drawings from the display screen.

F1 <HIST> — Draw the histogram.

[0, 100]4 by [-5, 5]1

Solution (D):

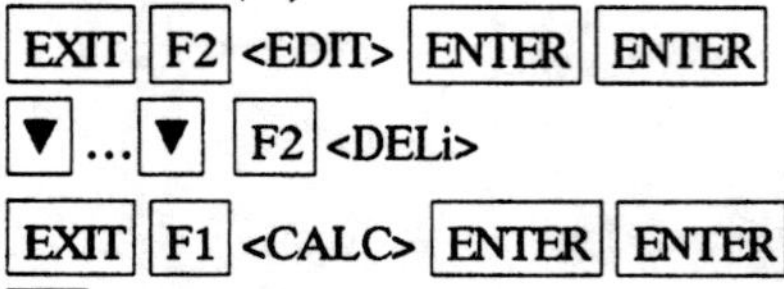

EXIT F2 <EDIT> ENTER ENTER — Get the data list and arrow down to the 89.

▼ ... ▼ F2 <DELi> — Delete the item.

EXIT F1 <CALC> ENTER ENTER — Get the statistics menu and the one variable sample statistics.

F1 <1-VAR> — The mean is now 39.5 and the sample standard deviation is 17.05. Notice how much influence an outlier has!

EXAMPLE 2: Given the following bivariate data: (A) find the correlation coefficient, (B) find the least squares line, (C) graph the least squares line, (D) plot the scatter diagram, and (E) sort the data on x and join the points using straight line segments.

x	5	8	10	29	9	24
y	3	4	8	20	3	19

Solution (A) & (B):

STAT F2 <EDIT> — Get the statistics menu.

ENTER ENTER F5 <CLRxy> 5 ENTER — Clear the statistical registers.

3 ENTER 8 ENTER 4 ENTER — Enter the data.

10 ENTER 8 ENTER etc. EXIT

F1 <CALC> ENTER ENTER — Linear regression. r = .98

F2 <LINR> — a=-1.72 and b=.79
So the least squares line is $\hat{y}$ = .79x - 1.72.

Solution (C):

EXIT EXIT

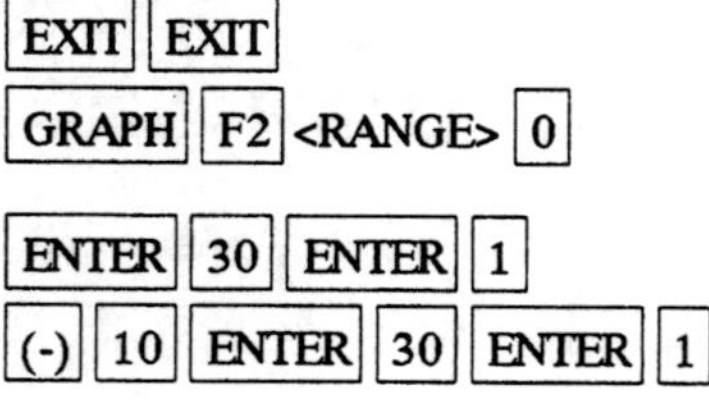

GRAPH F2 <RANGE> 0 — Set the graph RANGE. Observe the data to determine the RANGE values. Set the y value lower so if you trace, the coordinates at the bottom of the screen do not cover up the graph.

ENTER 30 ENTER 1

(-) 10 ENTER 30 ENTER 1

[F1] <y(x)=> — Deselect (or clear) all functions on the function list.

[F5] <SELCT> [EXIT] [EXIT] — See Example 1 above.

[STAT] [F3] <DRAW> [F4] <DRREG> — Draw the regression line that you just calculated.

Solution (D):

[F2] <SCAT> — The scatter diagram will draw on the same graph as the least squares line.

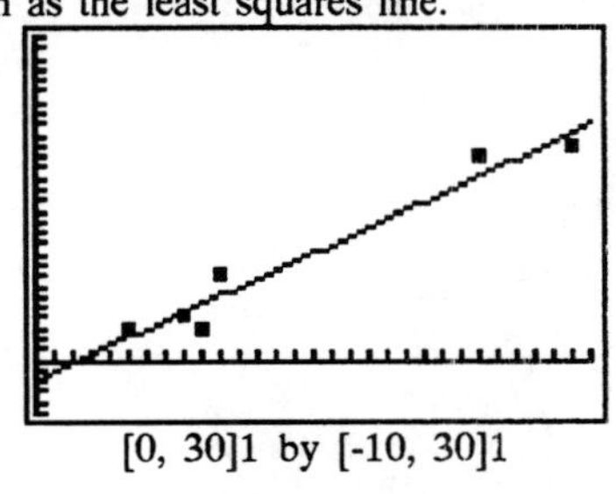

[0, 30]1 by [-10, 30]1

Solution (E):

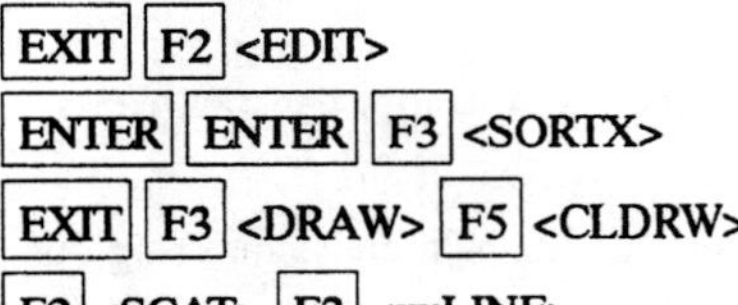

[EXIT] [F2] <EDIT>

[ENTER] [ENTER] [F3] <SORTX> — Sort the data on x variable.

[EXIT] [F3] <DRAW> [F5] <CLDRW> — Clear the drawing. Draw the scatter diagram.

[F2] <SCAT> [F3] <xyLINE> — Draw the line joining the points. They will join in the order of the x data in the data list.

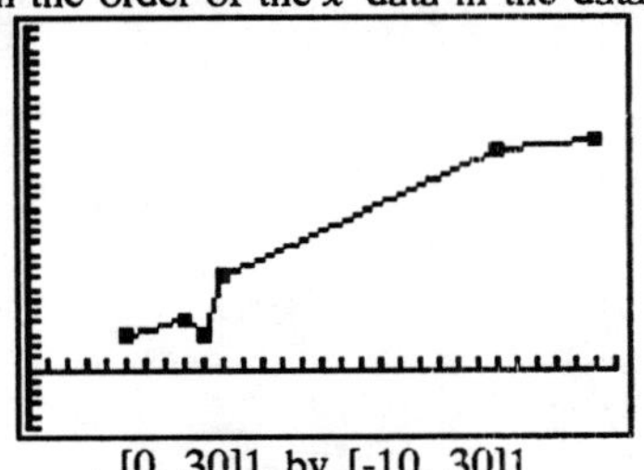

[0, 30]1 by [-10, 30]1

C-17 Setting the Display Mode for Numbers

Numbers can be entered into the calculator in scientific notation, a set number of significant digits, or a fixed number of decimal places.

EXAMPLE 1: Calculate $(9.35417\times 10^{4})(-3.64389\times10^{-6})$ using:
(A) scientific notation, (B) two significant digits, and (C) Four decimal place accuracy.

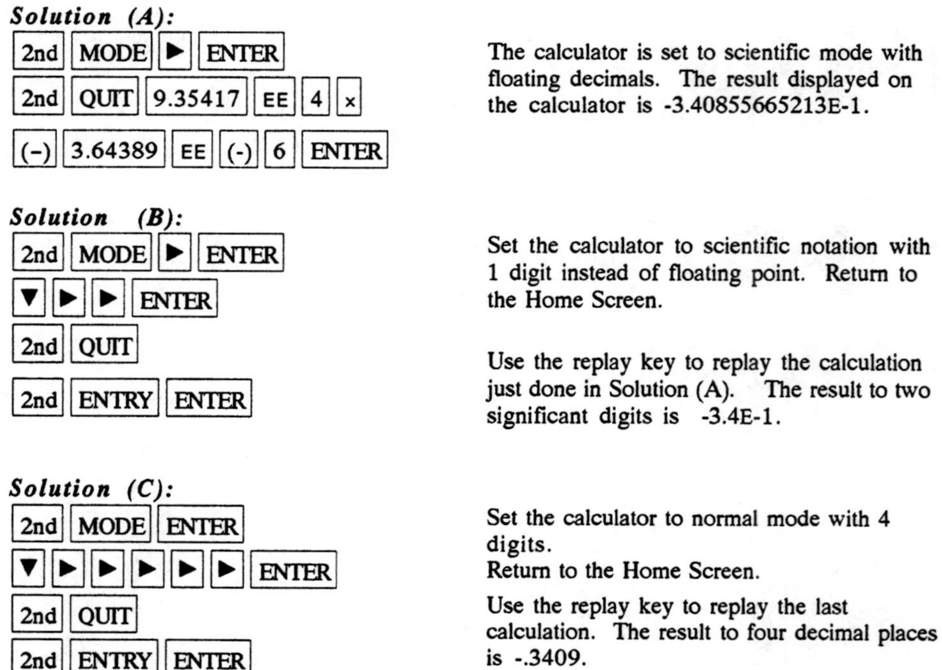

Solution (A):

2nd MODE ▶ ENTER

2nd QUIT 9.35417 EE 4 ×

(-) 3.64389 EE (-) 6 ENTER

The calculator is set to scientific mode with floating decimals. The result displayed on the calculator is -3.40855665213E-1.

Solution (B):

2nd MODE ▶ ENTER

▼ ▶ ▶ ENTER

2nd QUIT

2nd ENTRY ENTER

Set the calculator to scientific notation with 1 digit instead of floating point. Return to the Home Screen.

Use the replay key to replay the calculation just done in Solution (A). The result to two significant digits is -3.4E-1.

Solution (C):

2nd MODE ENTER

▼ ▶ ▶ ▶ ▶ ▶ ENTER

2nd QUIT

2nd ENTRY ENTER

Set the calculator to normal mode with 4 digits.
Return to the Home Screen.

Use the replay key to replay the last calculation. The result to four decimal places is -.3409.

Change the calculator back to floating point. The examples in this manual are done using this setting unless otherwise noted.

NOTES

Appendix D

TI-86 Operations

D-1 Operations

Press ON to turn on the calculator.

Press CLEAR to clear the screen.

Press EXIT to clear any menus displayed at the bottom of the screen.

Press 2nd ▲ to make the display darker.

Press 2nd ▼ to make the display lighter.

Press 2nd 3 to get the RESET screen.

Then press F3 <RESET> F1 <ALL> to clear all the memories, and press F4 <YES> to clear. You must now adjust the display so you can see it. Enter a few numbers and adjust the display contrast as described above.

RAM	DELET	RESET	TOL	ClrEnt
ALL	MEM	DFLTS		

Mem Cleared
Defaults set

Press 2nd OFF to turn off the calculator.

Home Screen

The screen where calculations are done is called the Home Screen. You can always get to the Home Screen by pressing 2nd QUIT. Note: QUIT is above the EXIT key.

> The item above the key will be referenced as if it is on the key. For example, QUIT, which is above the EXIT key, will be referenced from now on as 2nd QUIT.

2nd

This key must be used to access the calculator functions above and to the left of a key cap. A new cursor is displayed, ↑. These functions are in yellow on the calculator face.

ALPHA and 2nd ALPHA

This key is used to access the calculator functions and memory locations above and to the right of a key cap. These functions are in light blue on the calculator face.

The cursor changes to an A when ALPHA is pressed. Pressing ALPHA and then a letter will display an upper case letter.

The cursor changes to a a when 2nd ALPHA is pressed. Pressing 2nd ALPHA will display a lower case letter.

NOTE: There is a separate memory location for the upper case letter and for the lower case letter.

MODE

This is found above the MORE key and is used to set the operating mode of the calculator. Press 2nd MODE to get the mode screen. When the highlighted items are on the left, the calculator is in default mode. To choose a different mode: Use the arrow keys to choose the item you want to change to, then press the ENTER key. The new selection will now be highlighted. Press 2nd QUIT to return to the Home Screen.

Menus

Menus are listed across the bottom of the screen. To choose a menu item press the function key immediately below the menu item. In this manual, these functions will be referred to as the key to press in a box followed by the function in angle brackets. For example, the trace function of the calculator is found on the GRAPH menu above the F4 key is denoted by F4 <TRACE>.

Error Correction

Use the 2nd INS key to insert characters. (INS is above the DEL key.)

Use the DEL key to delete characters.

Use the arrow keys ◄ or ► to move to the left or right.

Use the up arrow key ▲ to move up one line.

Use the down arrow key ▼ to move down one line.

2nd ENTRY is used to replay the last executed operation. (ENTRY is above the ENTER key.)

D-2 Calculating

Many calculator functions are found above key caps or in menus. For example the square root function is above the x^2 key but the absolute value is found in the [MATH] menu in the [NUM] submenu.

EXAMPLE 1: Calculate $|4^2 - \sqrt{34}\ |$.

Solution: [2nd] [MATH] [F1] <NUM> [F5] <abs> [(] [4] [x^2] [-] [2nd] [$\sqrt{\ }$] [34] [)] [ENTER].

[ENTER] tells the calculator to calculate. The result displayed is 10.1690481052.

D-3 Plotting Points and Drawing Line Segments

EXAMPLE 1: (A) Plot the points (-3, 6) and (1, 5). (B) Draw the line segment having these points as the endpoints.

Solution (A):

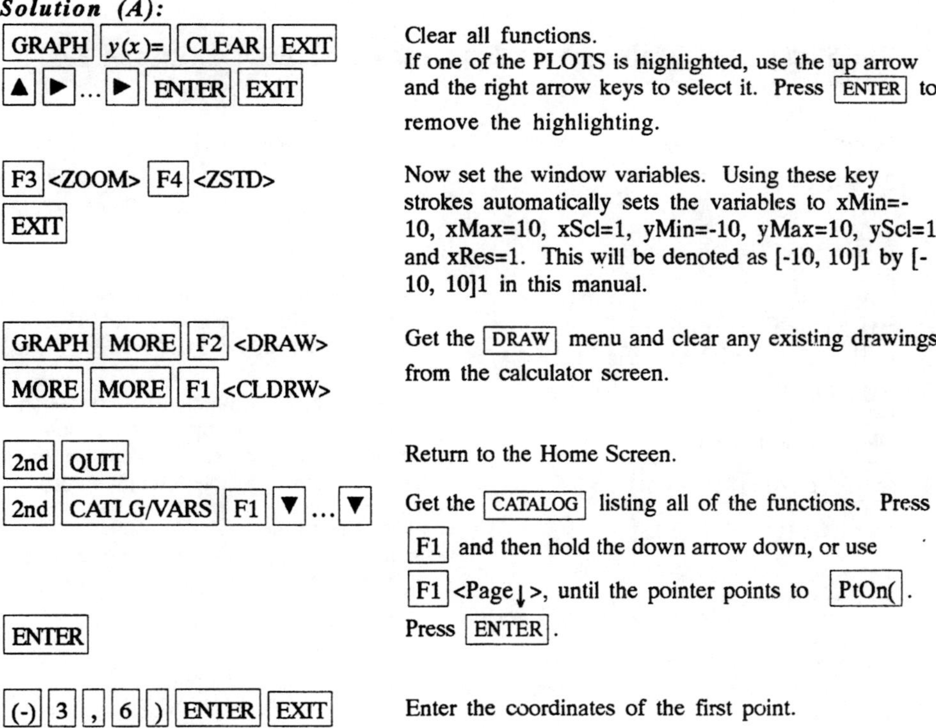

Keystrokes	Explanation
[GRAPH] [$y(x)=$] [CLEAR] [EXIT] [▲] [▶] ... [▶] [ENTER] [EXIT]	Clear all functions. If one of the PLOTS is highlighted, use the up arrow and the right arrow keys to select it. Press [ENTER] to remove the highlighting.
[F3] <ZOOM> [F4] <ZSTD> [EXIT]	Now set the window variables. Using these key strokes automatically sets the variables to xMin=-10, xMax=10, xScl=1, yMin=-10, yMax=10, yScl=1 and xRes=1. This will be denoted as [-10, 10]1 by [-10, 10]1 in this manual.
[GRAPH] [MORE] [F2] <DRAW> [MORE] [MORE] [F1] <CLDRW>	Get the [DRAW] menu and clear any existing drawings from the calculator screen.
[2nd] [QUIT]	Return to the Home Screen.
[2nd] [CATLG/VARS] [F1] [▼] ... [▼]	Get the [CATALOG] listing all of the functions. Press [F1] and then hold the down arrow down, or use [F1] <Page↓>, until the pointer points to [PtOn(].
[ENTER]	Press [ENTER].
[(-)] [3] [,] [6] [)] [ENTER] [EXIT]	Enter the coordinates of the first point.

Keystrokes	Explanation
2nd ENTRY	Use the replay feature of the calculator to get the last entry.
◄ ... ◄ DEL 1 , 5) ENTER	Use the left arrow and change the coordinates to those of the second point. Delete any unwanted characters using DEL .

Solution (B):

Keystrokes	Explanation
2nd QUIT	Return to the Home Screen.
2nd CATLG/VARS F1 ▼ ... ▼ ENTER	Get the CATALOG listing all of the functions. Press F1 and then the down or up arrow, or F1 <Page↓>, or F2 <Page↑>, until the pointer points to Line(. Press ENTER .
(-) 3 , 6 , 1 , 5) ENTER	Enter the coordinates of the first point and the second point separated by commas. Press ENTER to draw the line segment.

D-4 Storing and Evaluating an Expression, and Using Lists and Tables

EXAMPLE 1: Store the expression $\frac{-b+\sqrt{b^2-4ac}}{2a}$ and evaluate it to find one solution of $3x^2+4x-5=0$.

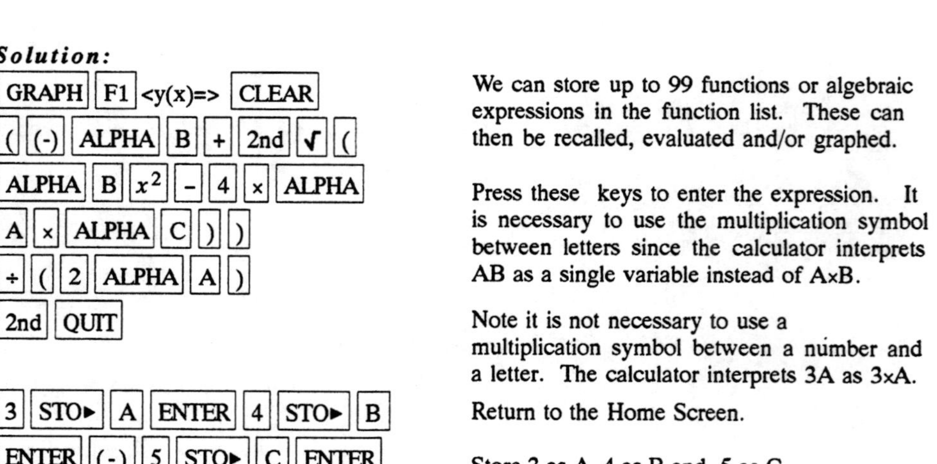

Solution:

Keystrokes	Explanation
GRAPH F1 <y(x)=> CLEAR	We can store up to 99 functions or algebraic expressions in the function list. These can then be recalled, evaluated and/or graphed.
((-) ALPHA B + 2nd √ (ALPHA B x^2 – 4 × ALPHA A × ALPHA C)) ÷ (2 ALPHA A)	Press these keys to enter the expression. It is necessary to use the multiplication symbol between letters since the calculator interprets AB as a single variable instead of A×B. Note it is not necessary to use a multiplication symbol between a number and a letter. The calculator interprets 3A as 3×A.
2nd QUIT	Return to the Home Screen.
3 STO► A ENTER 4 STO► B ENTER (-) 5 STO► C ENTER	Store 3 as A, 4 as B and -5 as C.

<u>METHOD 1</u>

Keystrokes	Comments
[2nd] [CATLG/VARS] [F2] <ALL> [F1] <PAGE↓>... [F1] <PAGE↓> [▼] [ENTER] [ENTER]	Get the variables list. Press [F1] <PAGE↓> and/or the arrow keys until the cursor is pointing at y1. Press [ENTER] [ENTER] The result is .786299647847.

<u>METHOD 2</u>

Keystrokes	Comments
[2nd] [ALPHA] [Y] [1] [ENTER]	Enter y1 on the screen and press [ENTER]. The result is .786299647847.

EXAMPLE 2: Given $f(x) = \sqrt[3]{x^4+2x}$ and $g(x) = 4^{2x-90}$. Find $f(x) + 3\sqrt{g(x)}$ at $x = 45.2$.

Solution:

Keystrokes	Comments
[GRAPH] [F1] <y(x)=> [CLEAR] [(] [x-VAR] [^] [4] [+] [2] [x-VAR] [)] [^] [(] [1] [÷] [3] [)] [ENTER] [CLEAR] [4] [^] [(] [2] [x-VAR] [−] [90] [)] [2nd] [QUIT]	Enter the functions as y1 and y2. Press [CLEAR] to delete the old y1 function. Note that lower case x is used in functions. Upper case and lower case letters are different. Values can be stored for each of these.
[45.2] [STO▸] [x-VAR] [ENTER]	Store 45.2 as x.
[2nd] [ALPHA] [Y] [1] [+] [3] [×] [2nd] [√] [(] [2nd] [ALPHA] [Y] [2] [)] [ENTER]	Form $f(x) + 3\sqrt{g(x)}$ as y1 + 3√ y2. The expression will be evaluated when you press [ENTER]. The result is 164.97 to two decimal places.

EXAMPLE 3: Evaluate the expression $\frac{3x^2 + 5x - 8}{x^2 - 1}$ to three decimal places for $x = .900$, .990, .999, 1.001, 1.010, and 1.100 by: (A) evaluating the expression for each value of x individually; and (B) using a list.

Solution (A): First store the expression in the Y= list. Then store the value for x. Recall the expression and evaluate it.

Keystrokes	Comments
[GRAPH] [F1] <y(x)=> [CLEAR] [(] [3] [x-VAR] [^] [2] [+] [5] [x-VAR] [−] [8] [)] [÷] [(] [x-VAR] [^] [2] [−] [1] [)] [2nd] [QUIT]	Clear the old expression for y1 by pressing [CLEAR]. Store the expression as y1 and return to the Home Screen.

Keystrokes	Description
.9 STO► x-VAR ENTER	Store the values of x.
2nd ALPHA Y 1 ENTER	Recall the expression and evaluate.

Store the next value for x and reevaluate the expression. Repeat for the other values of x. The corresponding results are: 5.632, 5.513, 5.501, 5.499, 5.488, and 5.381, rounded to three decimal places.

Solution (B): Store the values of x in a list and evaluate the expression for each value of x simultaneously.

Keystrokes	Description
GRAPH F1 <y(x)=> CLEAR	Clear the old expression for y1 by pressing CLEAR.
(3 x-VAR ^ 2 + 5 x-VAR − 8) ÷ (x-VAR ^ 2 − 1) 2nd QUIT	Store the expression as y1 and return to the Home Screen.
2nd LIST F1 <{> .900 , .990 , .999 , 1.001 , 1.010 , 1.100 F2 <}> STO► x-VAR ENTER	Store the values of x in the list x.
2nd alpha Y 1 ENTER	Simultaneously calculate the value of the expression stored as y1 for the values of the list x. The results are: 5.632, 5.513, 5.501, 5.499, 5.488, and 5.381, rounded to three decimal places. Use the right arrow key to view the other values in the list.

EXAMPLE 4: Evaluate the expression $\frac{3x^2 + 5x - 8}{x^2 - 1}$ to three decimal places for values of x at each tenth from 0 to 1 using a table.

Solution: First store the expression in the Y= list. Set the table parameters to begin at $x = 0$ and to have an increment of .1. Get the table.

Keystrokes	Description
GRAPH F1 <y(x)=> CLEAR	Clear the old expression for y1 by pressing CLEAR.
(3 x-VAR ^ 2 + 5 x-VAR − 8) ÷ (x-VAR ^ 2 − 1) 2nd QUIT	Store the expression as y1 and return to the Home Screen.

[TABLE] [F2] <TBLST> [0] [ENTER] [.1] [ENTER] — Set the table to begin evaluating the expression at x = 0 with a step size of .1.

[ENTER] — Set the calculator to automatically display values of x and y1.

[F1] <TABLE> [TABLE] [▼] ... [▼] — Get the table. Arrow down to see the rest of the table. The highlighted value will appear at the bottom of the table.

x	y1	
0	8	
.1	7.545455	
.2	7.166667	
.3	6.846154	
.4	6.571429	
.5	6.333333	
X=0		

D-5 Evaluating Inequalities

EXAMPLE 1: Given the inequality $x^2 - 3t < 5x + 9t^2$. Test this inequality for x = 1.89 and t = -4.52.

Solution:

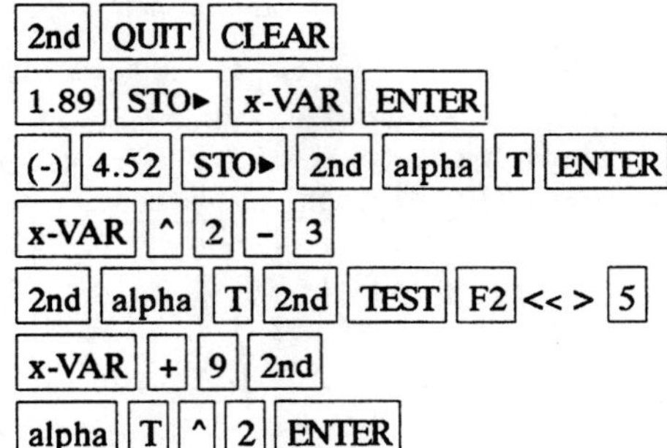

Store the value of the variables. Lower case letters are used.

Enter the expression and evaluate.

The result is 1. This means the inequality is true for these values of the variables. If the inequality was false, the result would have been 0.

D-6 Graphing and Style of Graph

EXAMPLE 1: Graph $y = 3x^2 - 6$ and $y = 3x^2 + 1$ on the same graphing screen. Use a solid line for the first graph and a dotted line for the second.

Solution:

[GRAPH] [MORE] [F2] <DRAW>

[MORE] [MORE] [F1] <CLDRW>

Clear drawings.

[EXIT] [EXIT]

Turn statistical plots off by pressing the up arrow, selecting the Plot, and then pressing [ENTER].

Clear or deselect all other functions by placing the cursor at the expression for the function and pressing [F5] <SELCT>. Now this deselected function will not graph. Note that the = sign is no longer highlighted.

[GRAPH] [F1] <y(x)=> [CLEAR] [3] [x-VAR] [^] [2] [-] [6] [MORE] [F3] <STYLE>... [F3] <STYLE>

Enter the first functions.

Select the style of graph by using the STYLE option. Press [F3] <STYLE> repeatedly until the solid line shows to the left of y1.

[ENTER] [CLEAR] [3] [x-VAR] [^] [2] [+] [1] [F3] <STYLE>... [F3] <STYLE>

Enter the second function and select the dotted style of graph by using [F3] <STYLE>.

[EXIT] [F3] <ZOOM> [F4] <ZSTD>

Exit the function menu by pressing [EXIT].

Press [F3] and then [F4] to get the standard viewing window. This will automatically set the window variables to have $-10 < x < 10$ and $-10 < y < 10$ and graph the functions.

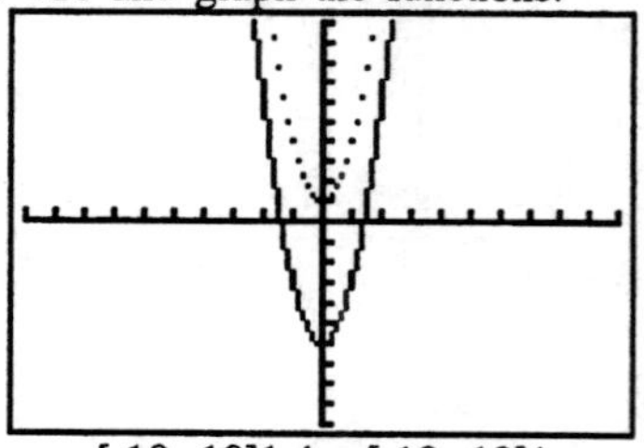

[-10, 10]1 by [-10, 10]1

[CLEAR]

[2nd] [QUIT]

Note that pressing [CLEAR] will remove the menus from the bottom of the screen.

> **[ZSTD]** automatically sets the viewing window variables to $-10 < x < 10$ and $-10 < y < 10$. The viewing window variables will be denoted in this manual by the notation [-10, 10]1 by [-10, 10]1. The window variables can be set by pressing [GRAPH] [F2] <WIND> and entering the desired values.

> **[F2] <WIND>** means to choose the WIND option from the menu on the screen above the [F2] key.
>
> The function will <u>not</u> graph if you choose a function on the [F1] <y(x)=> menu and then press [F5] <SELCT>. You will notice that the highlighting disappears from the = sign. You can still evaluate the expression. To select the function, press [F5] <SELCT> again.

EXAMPLE 2: Graph $f(x) = \begin{cases} .5x + 1 & x \le 2 \\ x^2 + 2 & x > 2 \end{cases}$.

Solution:
The calculator will only graph functions for values of x for which the function is defined, so we write this piecewise-defined function in a form where the pieces are defined only for the values of x that we want. Enter two functions as y1=$(.5x+1)\div(x\le2)$ and y2= $(x^2+2)\div(x>2)$. The first function is defined with $x \le 2$ and the second function is defined when $x > 2$.

[GRAPH] [F1] <y(x)=> [CLEAR] — Clear the existing function.

[(] [.5] [x-VAR] [+] [1] [)] [÷] [(] [x-VAR] [2nd] [TEST] [F4] <≤> [2] [)] [ENTER] — Enter the first piece of the piecewise-defined function.

[CLEAR] [(] [x-VAR] [^] [2] [+] [2] [)] [÷] [(] [x-VAR] [2nd] [TEST] [F3] <>> [2] [)] [ENTER] — Clear the existing function. Enter the second piece of the piecewise-defined function.

[EXIT] [EXIT] [F3] <ZOOM>

[F4] <ZSTD> — Graph.

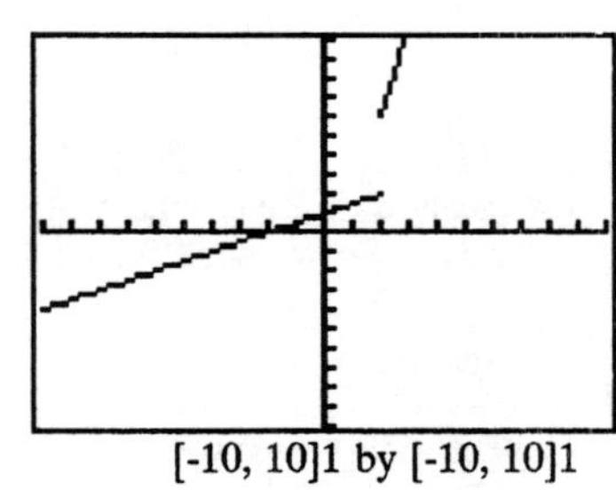

[-10, 10]1 by [-10, 10]1

D-7 TRACE, ZOOM, WINDOW, SOLVER, ROOT Finder and INTERSECT

EXAMPLE 1: Find the positive x intercept(s) for $y = 3x^2 - 6$. Use three decimal place accuracy.

There are six methods we can use to approximate the *x* intercepts.

1. We can trace the function and zoom in on the point by changing the window variable values.
2. We can trace the function and zoom in on the point by setting the Zoom Factors and using ZOOM.
3. We can zoom in on the point by using the Zoom Box feature and then tracing the function.
4. We can use the solver feature of the calculator.
5. We can use the root finder feature of the calculator.
6. We can use the intersect feature of the calculator.

Solution: Store the function as 3 X ^ 2 - 6 as y1. Graph using the [-10, 10]1 by [-10, 10]1. (see Example 1 from Section D-6).

You see a new cursor near the center of the graph screen and the coordinates of that point on the graph displayed at the bottom of the screen.

Use right and left arrow keys and move the cursor along the graph. If we had the graph of two or more functions displayed, we could use the up and down arrow keys to move between the functions. The 1 in the upper right-hand corner indicates that the cursor is on the graph of y1.

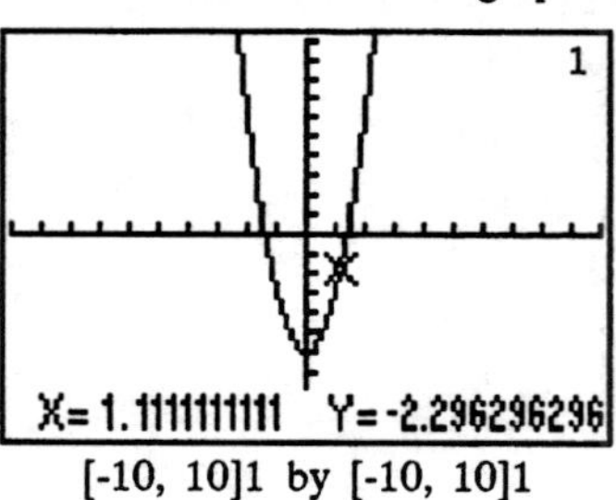

[-10, 10]1 by [-10, 10]1

METHOD 1 Set the window variables

Using the trace feature of the calculator, trace the function graphed above. We see the *x* intercept is near 1.5. So we shall set the window variables closer to that value.

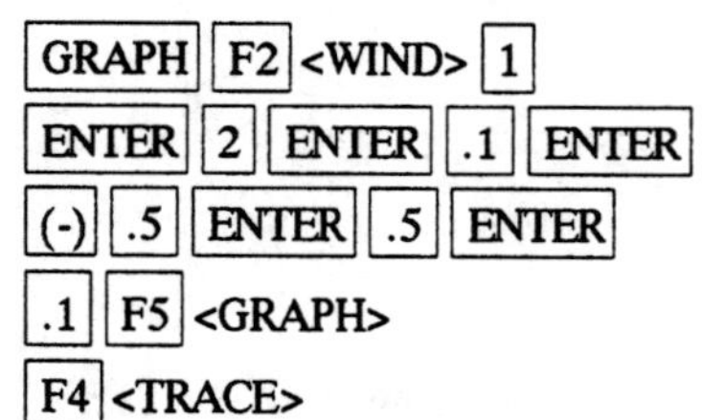

Set the window to [1, 2].1 by [-.5, .5].1.

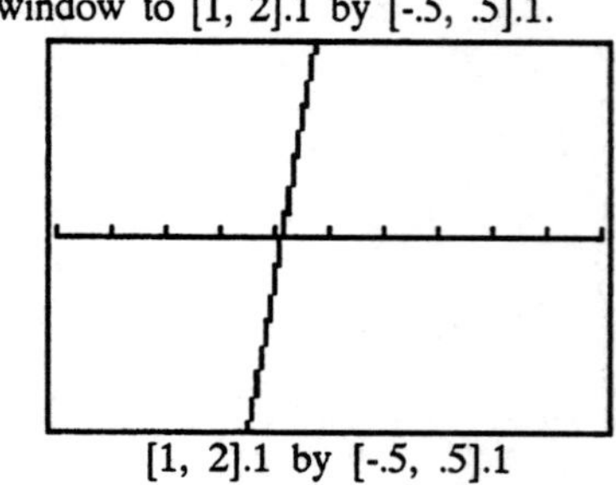

[1, 2].1 by [-.5, .5].1

Press EXIT to restore the menu at the bottom of the screen.

Use TRACE again to get a better approximation. The trace cursor always starts at the x value that is the middle of the window. It may not be on the display. Watch the values for x at the bottom of the screen.

We see that the x intercept is between 1.41 and 1.42. Set the window to [1.41, 1.42].01 by [-.01, .01].001 and graph again. We now find the x intercept to 1.414 accurate to three decimal places.

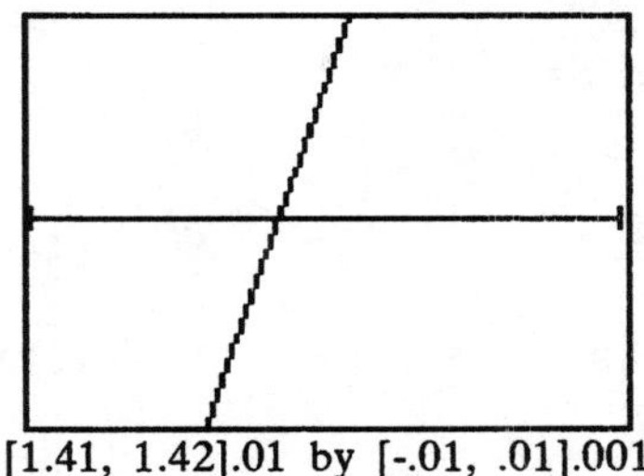

[1.41, 1.42].01 by [-.01, .01].001

METHOD 2 Zoom In

Keystrokes	Explanation
GRAPH F3 <ZOOM> F4 <ZSTD>	Get the [-10, 10]1 by [-10, 10]1 display as shown at the beginning of this section.
MORE MORE F2 <ZFACT> 5 ENTER 5	Set the Zoom Factors to 5 and 5. The graph will be 5 times as large in both the x and y directions.
F3 <ZOOM> F2 <ZIN>	
◄ or ► or ▲ or ▼	Use the ◄ or ► or ▲ or ▼ arrow keys to get the cursor to where you want the center of the new display, say at (1.42..., 0).
ENTER	The new display will be centered where you placed the cursor.

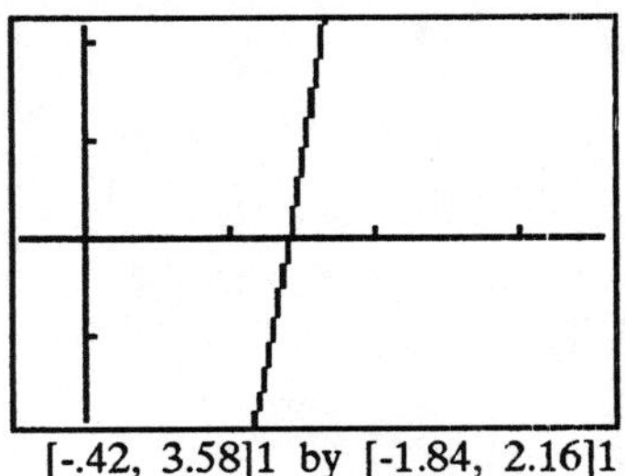

[-.42, 3.58]1 by [-1.84, 2.16]1

Keystrokes	Explanation
EXIT	Press EXIT to get the menu at the bottom of the screen. Use TRACE to get a new approximation.

Repeat the zooming in process as many times as necessary to get accuracy to three decimal places. The x intercept is 1.414 rounded to three decimal places.

METHOD 3 Zoom Box

[GRAPH] [F3] <ZOOM>

[F4] <ZSTD> — Get the [-10, 10]1 by [-10, 10]1 display as shown at the beginning of this section.

[F1] <BOX> — Get the Zoom Box option.

[◄] or [►] or [▲] or [▼] [ENTER]

[◄] or [►] or [▲] or [▼] [ENTER]

Use the arrow keys repeatedly to get the cursor close to but a little above and to the left of the x intercept, say at X = 1.1111111111, Y = .9677419355. Press [ENTER] to place the upper corner of the Zoom Box at this point. Now move the cursor down and to the right of the x intercept, say at X = 2.0634920635 Y = -.6451612903. Press [ENTER] again.

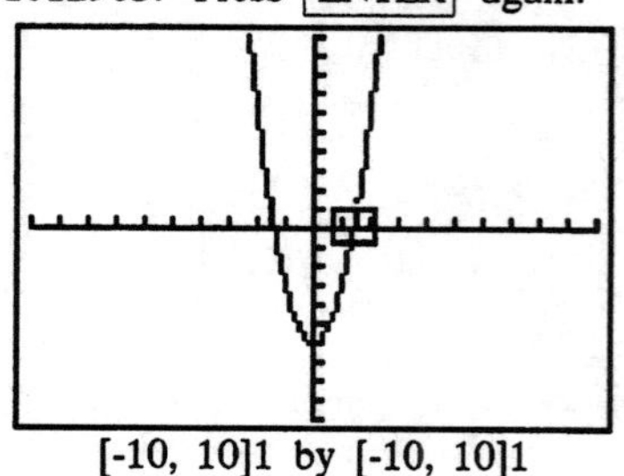

[-10, 10]1 by [-10, 10]1

The graph will be drawn again with the box as the outside of the new display.

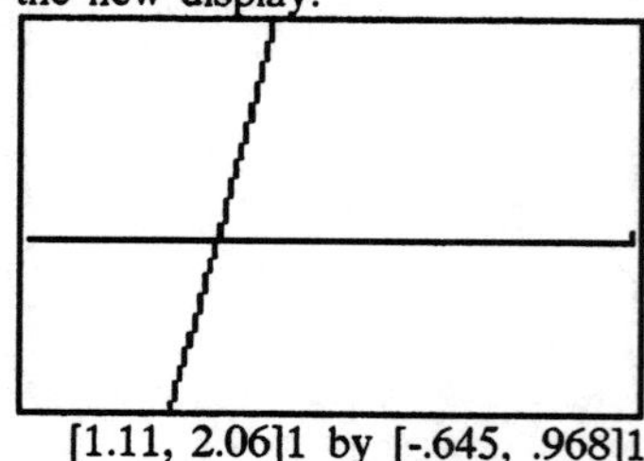

[1.11, 2.06]1 by [-.645, .968]1

Use Trace again to get another estimate of the x intercept. Use Zoom Box a second time to get a better approximation of the x intercept. The x intercept is 1.414 rounded to three decimal places.

METHOD 4 Solver

[2nd] [SOLVER] — Get the Solver feature.

[F1] <y1> [alpha] [=]

[0] [ENTER] [1.5] [F5] <SOLVE> — Enter an estimate of the x intercept and solve. The x intercept is 1.414, rounded to three decimal places.

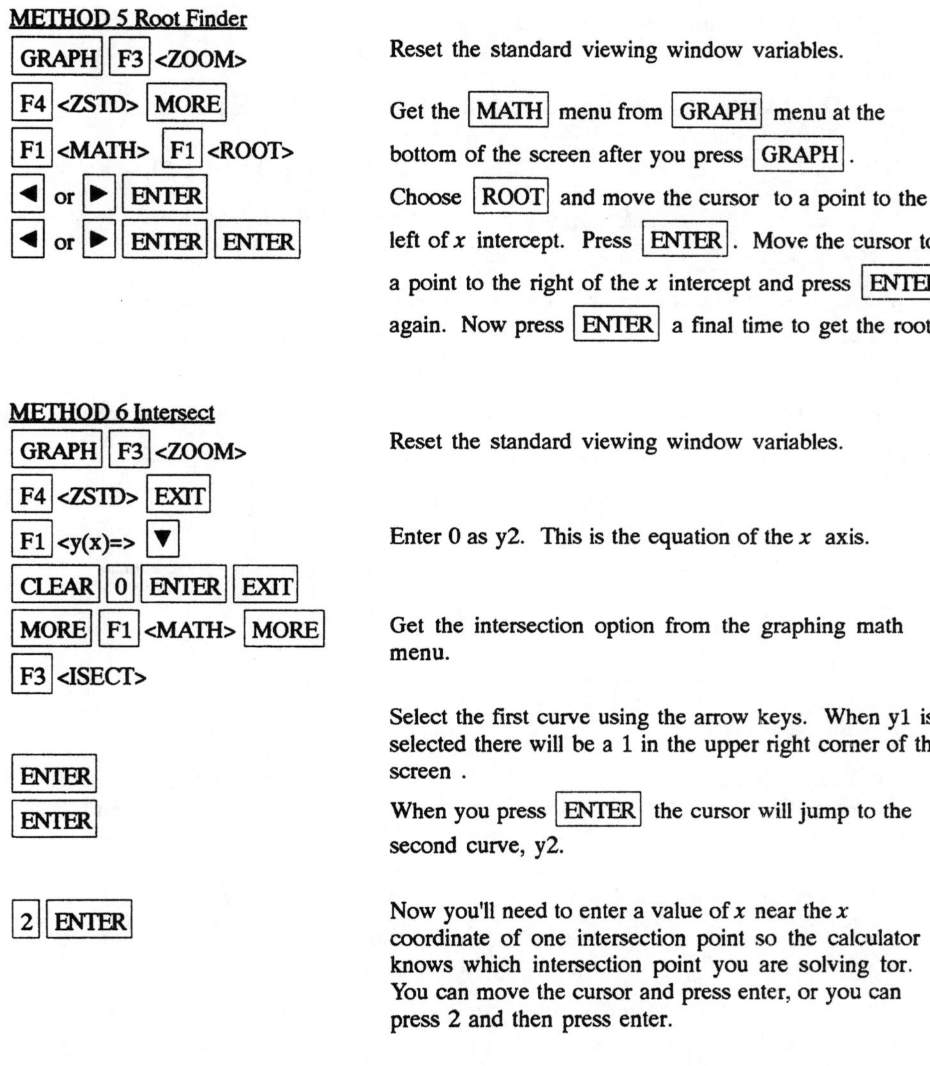

METHOD 5 Root Finder

Keystrokes	Explanation
GRAPH F3 <ZOOM>	Reset the standard viewing window variables.
F4 <ZSTD> MORE	Get the MATH menu from GRAPH menu at the bottom of the screen after you press GRAPH.
F1 <MATH> F1 <ROOT>	Choose ROOT and move the cursor to a point to the left of x intercept. Press ENTER. Move the cursor to a point to the right of the x intercept and press ENTER again. Now press ENTER a final time to get the root.
◀ or ▶ ENTER	
◀ or ▶ ENTER ENTER	

METHOD 6 Intersect

Keystrokes	Explanation
GRAPH F3 <ZOOM>	Reset the standard viewing window variables.
F4 <ZSTD> EXIT	
F1 <y(x)=> ▼	Enter 0 as y2. This is the equation of the x axis.
CLEAR 0 ENTER EXIT	
MORE F1 <MATH> MORE	Get the intersection option from the graphing math menu.
F3 <ISECT>	Select the first curve using the arrow keys. When y1 is selected there will be a 1 in the upper right corner of the screen .
ENTER	
ENTER	When you press ENTER the cursor will jump to the second curve, y2.
2 ENTER	Now you'll need to enter a value of x near the x coordinate of one intersection point so the calculator knows which intersection point you are solving tor. You can move the cursor and press enter, or you can press 2 and then press enter.
	The result is $x = 1.414$, $y = 0$, rounded to three decimal places.

D-8 Choosing the WINDOW Variables Values

There is no one right way to determine the window variables that should be used to display the graph of a function. Three methods to try are:

1. Graph using the default window variable values and then zoom out.
2. Analyze the features of the function and then choose the window variable values.
3. Evaluate the function at several values and then choose the values for the window variables.

EXAMPLE 1 Graph $y = .03x^3 + .02x^2 - 5x + 10$.

Solution:

METHOD 1 Graph using the default window variable values and then zoom out

Enter the function as .03 z ^ 3 + .02 z ^ 2 - 5 z + 10.

Press EXIT F3 <ZOOM> F4 <ZSTD> to get the [-10, 10]1 by [-10, 10]1 display.

Press F3 <ZOOM> MORE MORE F2 <ZFACT> and set the Zoom Factors to xFact = 4 and yFact=4. Press F3 <ZOOM> F3 <ZOUT>.

Move the cursor so it is where you want the center of the new screen to be, and press ENTER. The result below used the origin as the center of the new screen. The screen has been zoomed out to give us a better view of the graph. Note the double lines as the axes. This is because the Xscl and Yscl values are too close together. They do not change when zooming out.

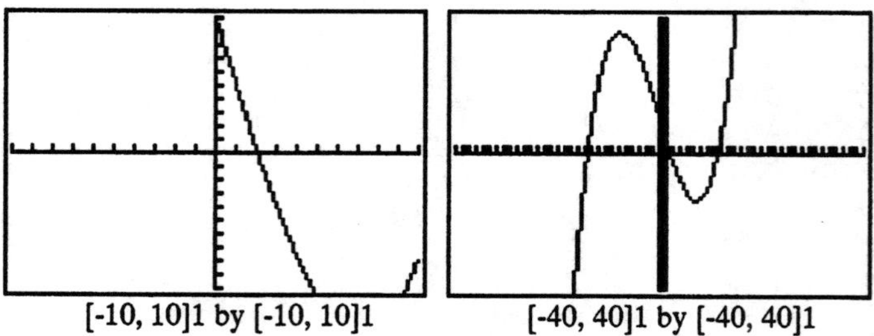

[-10, 10]1 by [-10, 10]1 [-40, 40]1 by [-40, 40]1

METHOD 2 Analyze the function and then choose the window variable values

We will analyze the term with highest power on x since this is the dominant term of the function.

We see that the leading coefficient is .03. This means that the vertical change in the graph will be approximately .03 unit for every time x^3 increases 1 unit. An equation that represents this is $.03x^3 = y$ or $x^3 = \frac{1}{.03}y$ or $x^3 \approx 30y$. So if y increases by 1 unit, x will increase by about $\sqrt[3]{30}$ or about 3 units. This indicates that we want the x axis three times as wide as the y axis. Hence, a good first choice for the horizontal axis is -30 to 30, and a good first choice for the vertical axis is -10 to 10. Set Xmin = -30, Xmax = 30 and Xscl = 3. We choose 3 as the Xscl because this will put 20 tick marks on the axis.

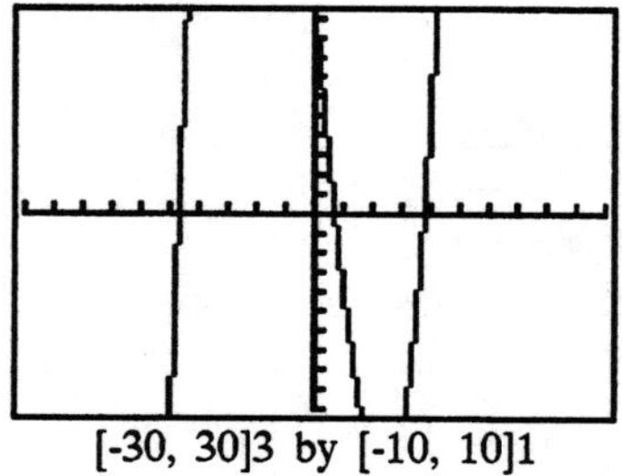

[-30, 30]3 by [-10, 10]1

The values for the vertical variables needs some adjusting since the turning points of the graph are not shown on the screen. Window variable values of [-30, 30]3 by [-40, 40]5 would show the turning points of the graph.

<u>METHOD 3 Evaluate the function at several values of x and then choose the window variable values</u>

Evaluate the function at several values of x and then choose the window variable values.The choice of x values to use when evaluating the function is arbitrary. Store a value of as x-VAR in the calculator and then evaluate the function. Repeat for all values of x.

x	$f(x)$
-25	-321.25
-10	32.00
0	10.00
10	-8.00
25	366.25

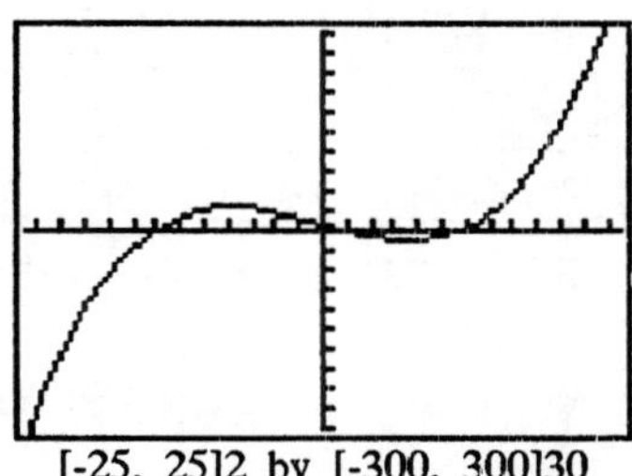

[-25, 25]2 by [-300, 300]30

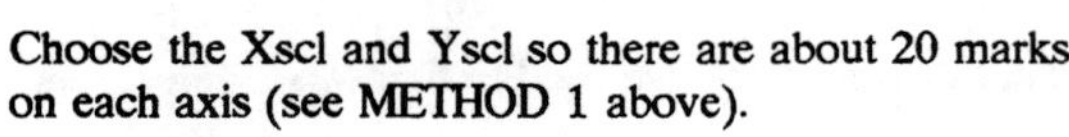

Choose the Xscl and Yscl so there are about 20 marks on each axis (see METHOD 1 above).

So a good first choice for the window variables could be [-25, 25]2 by [-300, 300]30.

Another way to evaluate the function is using the TABLE feature of the calculator (see Appendix Section D-4 of this manual). Set TblStart at -25, ΔTbl at 5, and the independent variable (Indpnt:) as Auto. We see the values range from -321.25 to 366.25 as we view the list for x values from -25 to 25.

D-9 Equations in One Variable

An equation in one variable can be solved graphically in two ways:

1. Algebraically move all terms to one side of the equation so that you have an expression equal to zero. Create a function $f(x)$ = (the expression). Graph the function.

   There are three ways to solve an equation in the form $f(x) = 0$ or to find the x intercepts of the graph of $f(x)$:

   a. Trace and zoom

   b. Use the solver

   c. Use the root finding feature

   d. Use the intersection point finder

   The set of x intercepts is the solution to the equation.

2. Create two functions: $f(x)$ = (the left side of the equation) and $g(x)$ = (the right side of the equation). Graph both functions. Trace and zoom, use the solver, or use ISECT to find the x value of the points of intersection of the two graphs. The set of x coordinates is the solution to the equation.

EXAMPLE: Solve $\sqrt[3]{x + 2} = 4x^2 - 3\sqrt{x+4}$. Use two decimal places.

Solution:

<u>METHOD 1</u> Write the equation as $\sqrt[3]{x + 2} - (4x^2 - 3\sqrt{x+4}) = 0$. Create the function $f(x) = \sqrt[3]{x + 2} - (4x^2 - 3\sqrt{x+4})$. Enter this function into the calculator and graph.

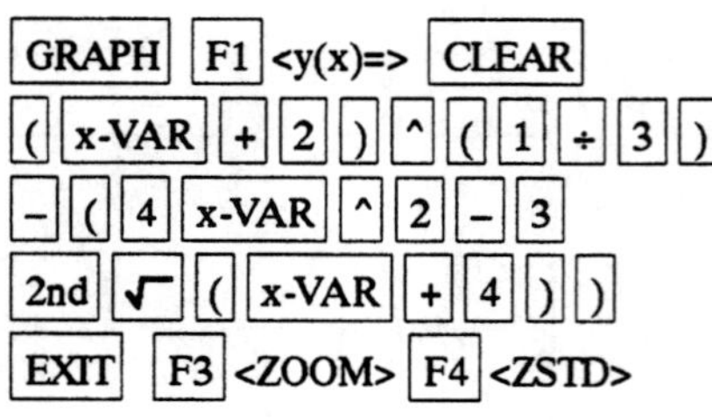

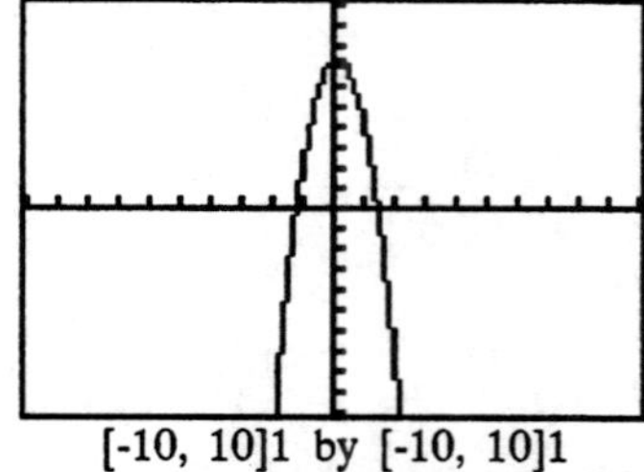

[-10, 10]1 by [-10, 10]1

We see the function crosses the x axis in two places. So we need to find two x intercepts.

Five methods for finding the x intercetps were described in Appendix Section D-7 of this manual. Following is the method using the Root Finder.

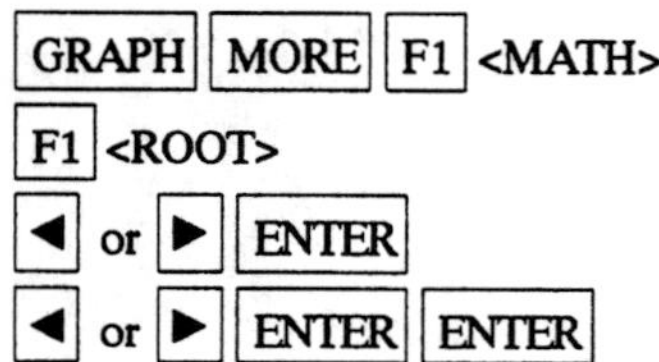

Get the MATH menu from GRAPH menu at the bottom of the screen after you press GRAPH. Choose ROOT and move the cursor to a point to the left of x intercept. Press ENTER. Move the cursor to a point to the right of the x intercept and press ENTER again. Now press ENTER a final time to get the root.

The two intercepts are $x = -1.22$ and $x = 1.46$. Hence the solutions to the equation are -1.22 and 1.46.

NOTE: To check for extraneous solutions, store the value in the calculator as x-VAR, recall the function, and evaluate it. (See Appendix Section D-4 of this manual).

<u>METHOD 2</u> Create the two functions $f(x) = \sqrt[3]{x + 2}$ and $g(x) = 4x^2 - 3\sqrt{x+4}$. Enter these into the function list as y1 and y2. Graph the functions and find the x coordinates of the intersection points. This can be done using trace and zoom, using the solver, or using the intersection point finder (see Appendix Section D-7 of this manual). The solver was used in the description below.

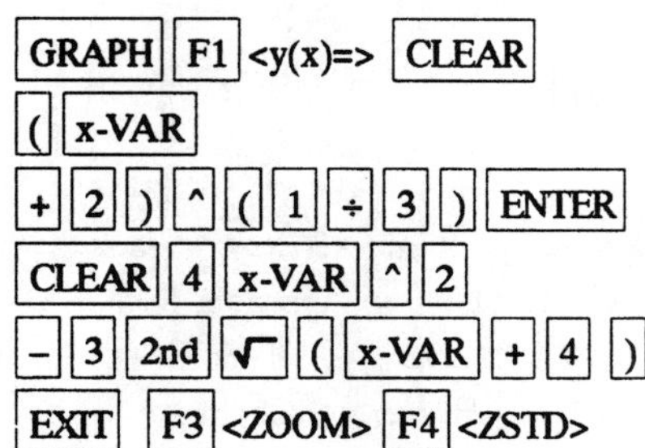

Clear the existing functions and enter the new ones.

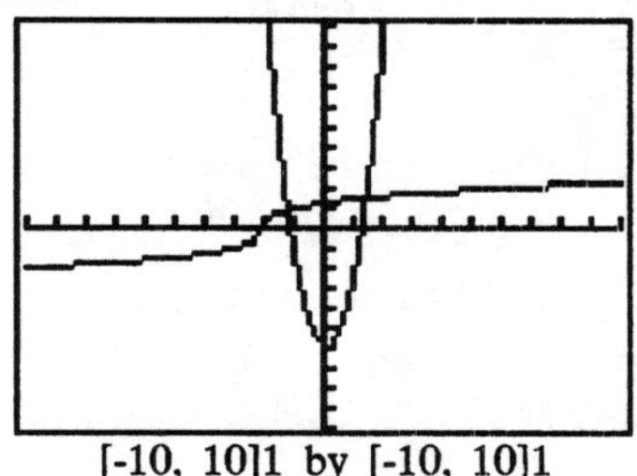

[-10, 10]1 by [-10, 10]1

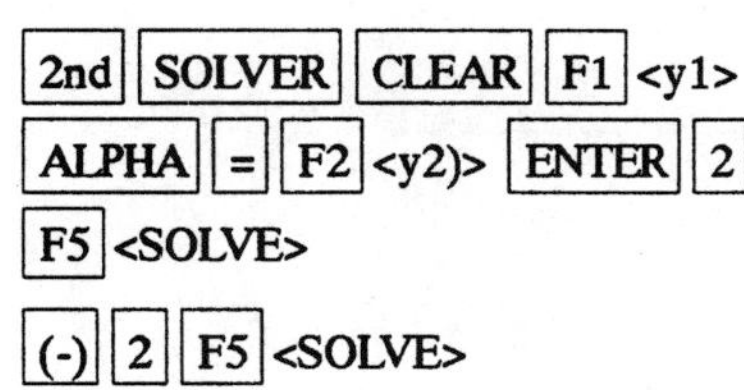

You'll need to enter a value of x near the x coordinate of one intersection point so the calculator knows which intersection point you are solving for. We chose 2. The result is $x = 1.46, y = 1.51$.

(-) 2 F5 <SOLVE>

Repeat for the other intersection point. We used -2 as the approximate value of this x coordinate. The result is $x = -1.22, y = .92$.

So the solution to this equation is $x = -1.22$ and $x = 1.46$.

D-10 Inequalities in One Variable

An inequality in one variable can be solved graphically in two ways:

1. Algebraically move all terms to one side of the equation so that you have an expression equal to zero. Create a function $f(x)$ = (the expression). Graph the function.

   There are several ways to solve an equation in the form $f(x) = 0$ (or to find the x intercepts of the graph of $f(x)$). See Appendix Section D-7 of this manual.

   The set of x intercepts are the cutoff values for the solution. Now determine the intervals over which the inequality is true and write the solution.

2. Create two functions: $f(x)$ = (the left side of the equation) and $g(x)$ = (the right side of the equation). Graph both functions. Find the x coordinate of the points of intersection of the two graphs (see Appendix Section D-8 Methods 2 of this manual). The set of x coordinates are the cutoff values for the solution. Determine the intervals over which the inequality is true and write the solution.

EXAMPLE: Solve $\sqrt[3]{x+2} < 4x^2 - 3\sqrt{x+4}$. Use two decimal places.

Solution:

<u>METHOD 1</u> Write the inequality as: $\sqrt[3]{x+2} - (4x^2 - 3\sqrt{x+4}) < 0$. Change the sign to an equal sign: $\sqrt[3]{x+2} - (4x^2 - 3\sqrt{x+4}) = 0$. Create the function $f(x) = \sqrt[3]{x+2} - (4x^2 - 3\sqrt{x+4})$, graph, and find the x intercepts. This was done above in Appendix Section D-9 Method 1 of this

manual. The cutoff values are $x = -1.22$ and $x = 1.46$. Observing the graph, we see it is below the x axis for $x < -1.22$ and $x > 1.46$. Hence the solution to this inequality can be written as $(-\infty, -1.22) \cup (1.46, \infty)$.

<u>METHOD 2</u> Create the two functions $f(x) = \sqrt[3]{x + 2}$ and $g(x) = 4x^2 - 3\sqrt{x+4}$. Graph these and find the x values of the intersection points. This was done above in Appendix Section D-9 Method 2 of this manual. The intersection points are (-1.22, .92) and (1.46, 1.51). Inspecting the graph we see that $f(x) < g(x)$ for $x < -1.22$ and $x > 1.46$. Hence the solution to this inequality can be written as $(-\infty, -1.22) \cup (1.46, \infty)$.

D-11 Permutations, Combinations, and Random Numbers

EXAMPLE 1: Find: (A) $P_{19,13}$ and (B) $C_{19,13}$.

Solution (A): This is a direct calculation.

Keystrokes	Explanation
[19] [2nd] [MATH] [F2] <PROB>	Enter the numbers and get the built-in function from the MATH menu. The result is displayed in scientific
[F2] <nPr> [13] [ENTER]	notation as 1.68951528346E14.
[EXIT] [EXIT]	To return to the Home Screen.

Solution (B): This is a direct calculation.

Keystrokes	Explanation
[19] [2nd] [MATH] [F2] <PROB>	Enter the numbers and get the built-in function from the MATH menu. The result is 27132.
[F3] <nCr> [13] [ENTER]	
[EXIT] [EXIT]	To return to the Home Screen.

EXAMPLE 2: Generate ten unique random numbers at least 0 and less than 50.

Solution:

Keystrokes	Explanation
[5] [STO▶] [2nd] [MATH] [F2] <PROB> [F4] <rand> [ENTER]	We need to store a "seed" number in the storage location RAND. This tells the random number generator where to start.
[EXIT] [F1] <NUM> [F2] <iPart> [(] [50] [×] [EXIT] [F2] <PROB> [F4] <rand> [)] [ENTER] [ENTER]	RAND generates a number greater than 0 and less than 1. If we multiply by 50 and take the integer part, the result is a number greater than or equal to 0 and less than 50.
[ENTER] [ENTER] [ENTER] [ENTER] [ENTER] [ENTER] [ENTER] [ENTER] [ENTER]	The 10 random numbers are: 36 13 6 17 21 ~~21~~ 20 5 31 16 48. Note that repetitions can occur but 21 cannot be used twice if unique random numbers are desired.

D-12 Matrices

Given the matrices $A = \begin{bmatrix} 1 & -4 & 6 \\ 2 & -1 & 0 \\ 5 & -3 & 6 \end{bmatrix}$ and $B = \begin{bmatrix} 4 & 15 \\ -2 & 6 \\ 2 & 5 \end{bmatrix}$.

EXAMPLE 1: Find -2.5AB.

Solution:

[2nd] [MATRX] [F2] <EDIT> [A] [ENTER] — Choose matrix A and enter the matrix dimensions.

[3] [ENTER] [3] [ENTER]

[1] [ENTER] [(-)] [4] [ENTER] [6] [ENTER] — Enter the matrix elements row by row.

[2] [ENTER] [(-)] [1] [ENTER] [0] [ENTER]

[5] [ENTER] [(-)] [3] [ENTER] [6] [EXIT]

[2nd] [MATRX] [F2] <EDIT> — Choose matrix B and enter the matrix dimensions.

[B] [ENTER]

[3] [ENTER] [2] [ENTER]

[4] [ENTER] [15] [ENTER]

[(-)] [2] [ENTER] [6] [ENTER] — Enter the matrix elements.

[2] [ENTER] [5]

[EXIT] — Return to the Home Screen.

[(-)] [2.5] [2nd] [MATRX] — Enter the matrix operations.

[F1] <NAMES> [F1] <A> [x] [F2] <B> [ENTER] — The result is $\begin{bmatrix} -60 & -52.5 \\ -25 & -60 \\ -95 & -217.5 \end{bmatrix}$.

EXAMPLE 2: Find A^{-1}.

Solution:

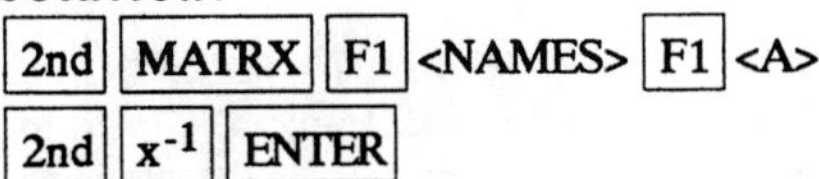

The operations are done from the Home Screen. The result rounded to three decimal places is:

$$A = \begin{bmatrix} -.167 & .167 & .167 \\ -.333 & -.667 & .333 \\ -.028 & -.472 & .194 \end{bmatrix}.$$

When performing operations on matrices, some numbers may look like decimals when in fact they are zero. This happens because the calculator rounds numbers to a certain number of decimal places. Examining the numbers using the arrow keys you will find they are decimals that are very close to zero. This situation can be eliminated by setting the number of decimal places to a specified number, say 3 (see Appendix Section D-17 of this manual).

EXAMPLE 3: Solve the system of equations using the inverse of the coefficient matrix. Round answers to two decimal places.

$x - 4y + 6z = 2.5$

$2x - y = 3.2$

$5x - 3y + 6z = -3.4$

Solution: The coefficient matrix was entered in Appendix Section D-12 Example 2 of this manual. Enter the constants as matrix B and find $A^{-1}B$.

[2nd] [MATRX]

[F2] <EDIT> [B] [ENTER] — Enter the matrix dimensions and enter the matrix elements.

[3] [ENTER] [1] [ENTER] [2.5] [ENTER]

[3.2] [ENTER] [(-)] [3.4] [EXIT] — Return to the Home Screen.

[2nd] [MATRX] [F1] <NAMES> — Perform the matrix operations.

[F1] <A> [2nd] [x^{-1}] [x] [F2] <B> [ENTER] — The result, rounded to two decmal places, is $\begin{bmatrix} -.45 \\ -4.10 \\ -2.24 \end{bmatrix}$. Hence, the solution to the system of equations is $x = -.45, y = -4.10$, $z = -2.24$, rounded to two decimal places.

EXAMPLE 4: Find the solution to the system of Example 3 above using Gauss-Jordan reduction. In other words, find the reduced form of the augmented coefficient matrix $\begin{bmatrix} 1 & -4 & 6 & 2.5 \\ 2 & -1 & 0 & 3.2 \\ 5 & -3 & 6 & -3.4 \end{bmatrix}$. Use two decimal places.

Solution:

METHOD 1 Use row operations on the matrix to get the reduced form.

[2nd] [MATRX] [F2] <EDIT> — Enter the matrix dimensions.

[A] [ENTER] [3] [ENTER] [4] [ENTER]

[1] [ENTER] [(-)] [4] [ENTER] [6] — Enter the matrix elements. Notice that if you are continuing from Example 1 - 3 of this section, you need only enter the elements in column 4. Use the down arrow to move from one row to the next.

[ENTER] [2.5] [ENTER] [2] [ENTER]

[(-)] [1] [ENTER] [0] [ENTER] [3.2]

[ENTER] [5] [ENTER] [(-)] [3] [ENTER]

[6] [ENTER] [(-)] [3.4] [EXIT] [EXIT]

[2nd] [MATRX] [F4] <OPS>
[MORE] [F5] <mRAdd>
[(-)] [2] [,] [EXIT] [F1] <NAMES>
[F1] <A> [,] [1] [,] [2] [)] [ENTER]
[STO▸] [F1] <A>

We wish to multiply by -2, matrix A, row 1 and add it to row 2. The result will be put in row 2.

`mRadd(-2,A,1,2)`

The result is $\begin{bmatrix} 1 & -4 & 6 & 2.5 \\ 0 & 7 & -12 & -1.8 \\ 5 & -3 & 6 & -3.4 \end{bmatrix}$.

Store the result in matrix A.

[ENTER] [2nd] [MATRX]
[F4] <OPS> [MORE] [F5] <mRAdd>
[(-)] [5] [,] [EXIT] [F1] <NAMES>
[F1] <A> [,] [1] [,] [3] [)] [ENTER]

Now multiply row 1 by -5 and add it to row 3 putting the result in row 3.

`mRadd(-5,A,1,3)`

The result is $\begin{bmatrix} 1 & -4 & 6 & 2.5 \\ 0 & 7 & -12 & -1.8 \\ 0 & 17 & -24 & -15.9 \end{bmatrix}$.

[STO▸] [A] [ENTER]

Store the result in matrix A.

[2nd] [MATRX] [F4] <OPS> [MORE]
[F4] <multR> [1] [÷] [7] [,] [ALPHA] [A]
[,] [2] [)] [ENTER]
[STO▸] [A]

Multiply row 2 by 1/7 .

The result is $\begin{bmatrix} 1 & -4 & 6 & 2.5 \\ 0 & 1 & -1.71 & -.26 \\ 0 & 17 & -24 & -15.9 \end{bmatrix}$

rounded to two decimal places.
Store the result in matrix A.

Continue in this manner until you get

$\begin{bmatrix} 1 & 0 & 0 & -.45 & \ldots \\ 0 & 1 & 0 & -4.1 & \ldots \\ 0 & 0 & 1 & -2.241666666 & \ldots \end{bmatrix}$.

The solution to the system of equations, rounded to two decimal places, is
$x = -.45, y = -4.10, z = -2.24$.

> Note: To swap rows of a matrix use [F4] <OPS> [MORE] [F2] <rSwap> from the [MATRX] menu. The command rSwap(A,2,3) will swap rows 2 and 3 in matrix A.

METHOD 2 Use the built-in capabilities of the calculator to get the reduced form.

[2nd] [MODE] [▼] [▶] [▶] [▶] [ENTER]

Set the mode to two decimal places.

[2nd] [MATRX] [F2] <EDIT>

Enter the matrix dimensions.

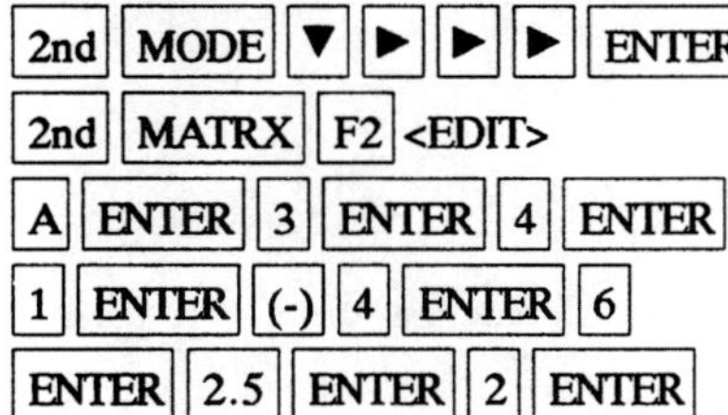

[A] [ENTER] [3] [ENTER] [4] [ENTER]
[1] [ENTER] [(-)] [4] [ENTER] [6]
[ENTER] [2.5] [ENTER] [2] [ENTER]

Enter the matrix elements. Notice that if you are continuing from Example 1 - 3 of this section, you need only enter the elements in column 4. Use the down arrow to move from one row to the next.

(-) 1 ENTER 0 ENTER 3.2 ENTER
5 ENTER (-) 3 ENTER
6 ENTER (-) 3.4 EXIT
2nd MATRX F4 <OPS> F5 <rref>
EXIT F1 <NAMES>
F1 <A> ENTER

We wish to get the reduced form of this matrix. Use the matrix operation rref to get this on the calculator.

The result is $\begin{bmatrix} 1.00 & 0.00 & 0.00 & -.45 \\ 0.00 & 1.00 & 0.00 & -4.10 \\ 0.00 & 0.00 & 1.00 & -2.24 \end{bmatrix}$.
Use the right and left arrow keys to view the entire matrix.

Hence, the solution to the system of equations, rounded to two decimal places, is $x = -.45, y = -4.10, z = -2.24$.

D-13 Logarithmic, Exponential and Hyperbolic Functions

EXAMPLE 1: Graph $f(x) = 10^{3.2x}$ for $-.5 < x < .5$.

Solution: Set the window variables to [-.5,.5].1 by [-10,10]1.

GRAPH F1 <y(x)=> CLEAR — Clear all functions from the function list.
10 ^ (3.2 x-VAR) EXIT
F2 <WIND> (-) .5 ENTER
.5 ENTER .1 ENTER (-) 10 ENTER
10 ENTER
1 F5 <GRAPH>

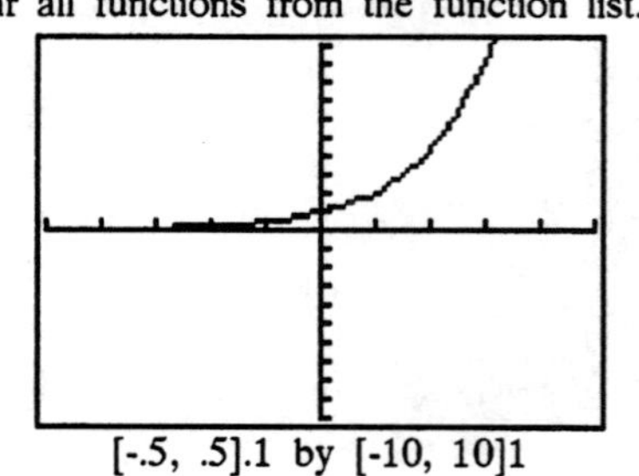

[-.5, .5].1 by [-10, 10]1

EXAMPLE 2: Graph $y = \dfrac{e^{x} + e^{-x}}{2}$.

Solution: This could be entered using (2nd e^x x-VAR + 2nd e^x (-) x-VAR) ÷ 2 . However the built-in hyperbolic cosine function can also be used.

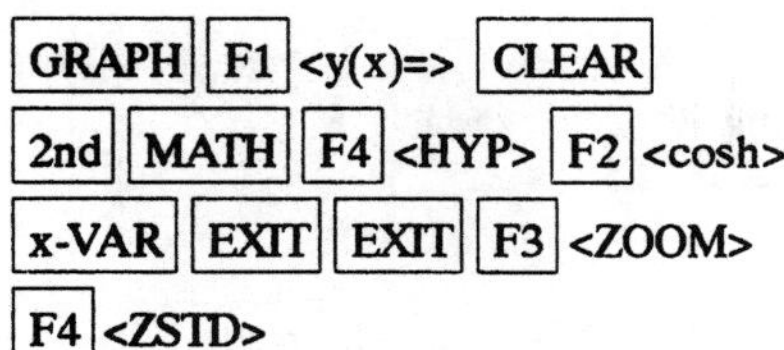

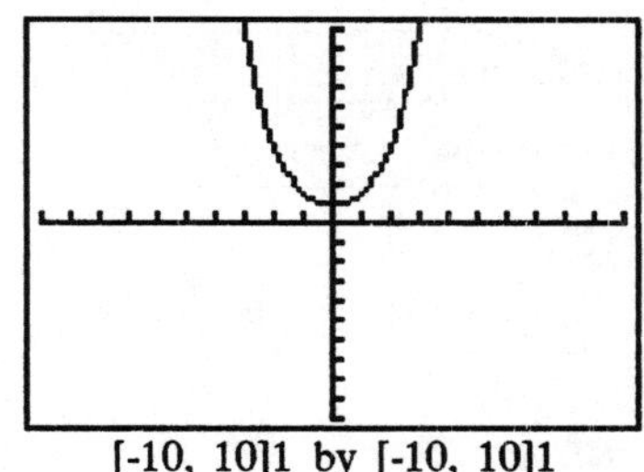

[-10, 10]1 by [-10, 10]1

EXAMPLE 3: Solve $\ln(x^2 + x) = 3x + 2$

Solution:

GRAPH F1 <y(x)=> CLEAR
LN (x-VAR ^ 2 + x-VAR
) ENTER CLEAR 3
x-VAR + 2 EXIT

Enter the function $f(x) = \ln(x + x^2)$ as y1 in the function list.

Enter the function $g(x) = 3x + 2$ as y2 in the function list.

F3 <ZOOM>
F4 <ZSTD>

Graph them on the same graphing screen. Find the x value of the point of intersection (see Appendix Section D-7 Method 6 of this manual). This will be the solution to the equation.

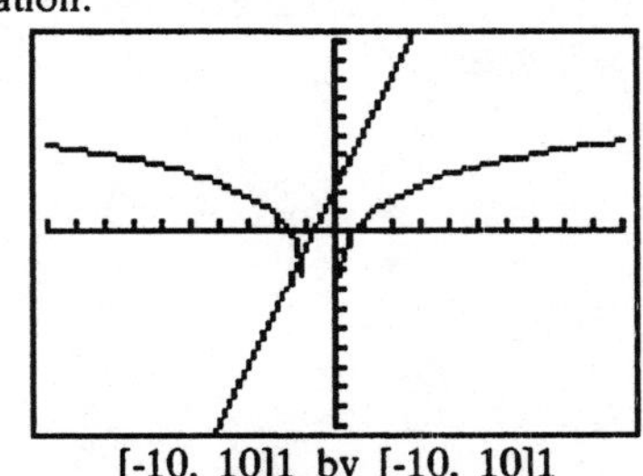

[-10, 10]1 by [-10, 10]1

The solution is -1.18, rounded to two decimal places.

D-14 Trigonometric Functions

EXAMPLE 1: Evaluate $f(x) = \cos x + \csc x$ at (A) 53°45'18" and (B) 78.52 radians.

Solution: (A)

GRAPH F1 <y(x)=> CLEAR
COS x-VAR + 1 ÷ SIN
x-VAR EXIT EXIT

Clear or deselect all existing functions. Store the function as y1.

2nd MODE ▼ ▼ ▶ ENTER

Change the calculator to degree angle measure.

Keystrokes	Explanation
CLEAR 53 + 45 ÷ 60 + 18 ÷ 3600 STO▸ x-VAR ENTER	Change the angle to decimal degrees while storing it as the variable x. Store 53.755° as x.
2nd ALPHA y 1 ENTER	Evaluate the function. The result to two decimal places is 1.83.

Solution:* *(B)

Keystrokes	Explanation
2nd MODE ▼ ▼ ENTER	Change the calculator to radian angle measure.
2nd QUIT	Since the function was entered in part (A) above, we need only to store the angle measure and evaluate the stored function.
78.52 STO▸	Store 78.52 as x.
x-VAR ENTER	Recall the function. The result to two decimal places is 49.47.
2nd ALPHA y 1 ENTER	

EXAMPLE 2: Graph $f(x) = \cos x + \csc x$.

Solution: The function was entered in the calculator in Example 1 above.

Keystrokes	Explanation
GRAPH ZOOM MORE	Change the graph mode to Dot. (See Appendix Section D-6 of this manual.) This will eliminate the "vertical lines."
F3 <ZTRIG>	This option automatically sets the window variables to [-8.24,8.24]1.57 by [-4,4]1.

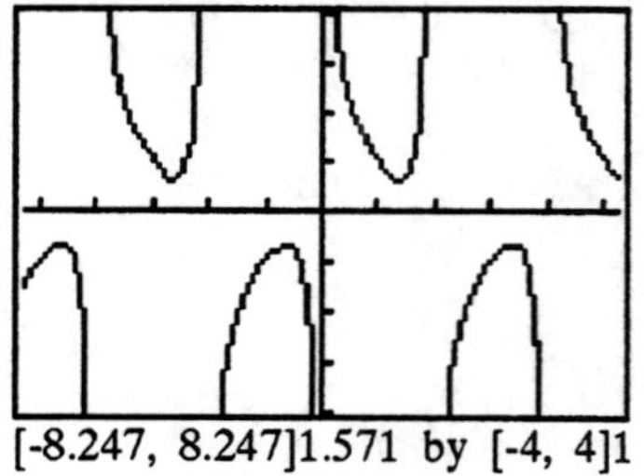

[-8.247, 8.247]1.571 by [-4, 4]1

D-15 Numeric Differentiation and Integration, Tangent Lines, and Differential Equations

EXAMPLE 1: (A) Evaluate $f(x) = \sqrt{x^5 - \sqrt[3]{x}}$ at $x = 3.54$. (B) Find the equation of the tangent line to $f(x)$ at 3.54. Use two decimal place accuracy. (C) Sketch the function and its tangent line using [2, 5]1 by [0, 60]5.

Solution (A): The function and it's derivative can be evaluated at this value of x .

Keystrokes	Explanation
[GRAPH] [F1] <y(x)=> [CLEAR]	Clear or deselect all existing functions. Enter the function and store it as y1.
[2nd] [√] [(] [x-VAR] [^] [5] [-]	
[x-VAR] [^] [(] [1] [÷] [3] [)] [)]	
[2nd] [QUIT]	Return to the Home Screen and store the value of x.
[3.54] [STO▸] [x-VAR] [ENTER]	
[2nd] [ALPHA]	Recall the function and press [ENTER] to
[Y] [1] [ENTER]	evaluate the function. $f(3.54) = 23.55$.

Soltuion (B):

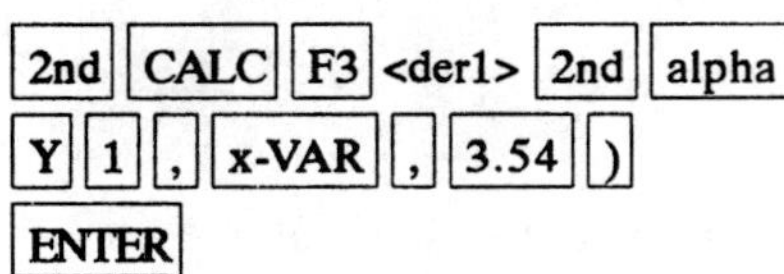

[2nd] [CALC] [F3] <der1> [2nd] [alpha]
[Y] [1] [,] [x-VAR] [,] [3.54] [)]
[ENTER]

Get the numeric derivative built-in function:
`der1(y1,x,3.54)`
This will evaluate the derivative with respect to x of the function stored in y1 at $x = 3.54$. The result is 16.67. (The default step size used in x increments is .001.)

The equation of the tangent line is $y - 23.55 = 16.67(x - 3.54)$.

Solution (C):

<u>Method 1</u> Enter the function as y1 in the calculator and the equation of the tangent line as y2 in the calculator and graph.

<u>Method 2</u> Enter the function as y1 in the calculator.
Set the viewing rectangle to [2, 5]1 by [10, 60]5.

Use the tangent drawing feature of the calculator by entering

[GRAPH] [MORE] [F2] <DRAW> [MORE]
[MORE] [MORE] [F2] <TanLn>
[2nd] [ALPHA] [Y] [1] [,] [3.54] [)] [ENTER].

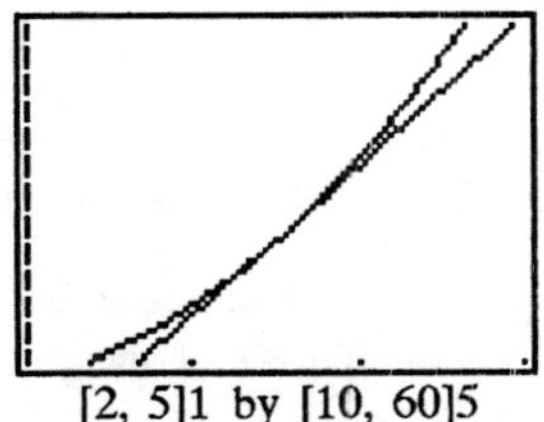

[2, 5]1 by [10, 60]5

(It may take a little while to graph.) Press [CLEAR] to remove the menu at the bottom of the screen. Press [EXIT] to display it again. The function and its tangent will be graphed as shown.

EXAMPLE 2: Is the function $f(x) = \sqrt{x^5 - \sqrt[3]{x}}$ concave upwards at $x = 3.54$?

Solution:

Enter the function and store it as y1 as shown in Example 1 of this section.

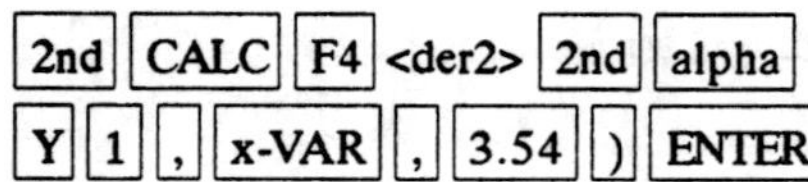

Use the second derivative function.

```
der2(y1,x,3.54)
```

The result is positive. Hence the function is concave upwards at $x = 3.54$.

EXAMPLE 3: Find $\int_{1.32}^{2.58} \sqrt{x^5 - \sqrt[3]{x}}\, dx$. Use two decimal place accuracy.

Solution:

[GRAPH] [F1] <y(x)=> [CLEAR]
[2nd] [√] [(] [x-VAR] [^] [5]
[−] [x-VAR] [^] [(] [1] [÷] [3]
[)] [)] [2nd] [QUIT]

Clear or deselect existing functions. Enter the function and store it.

[2nd] [CALC] [F5] <fnInt> [2nd] [alpha]
[y] [1] [,] [x-VAR] [,] [1.32] [,] [2.58] [)]
[ENTER]

Get the numeric integration function built into the calculator function:

```
fnInt(y1,x,1.32,2.58)
```

This will evaluate the definite integral. The result is 6.95. The default tolerance in calculating the definite integral is .00001.

This can be changed by pressing

[2nd] [MEM] [F4] <TOL>

EXAMPLE 4: Given the first-order differential equation $y' = 7y + 7x - 3$ with $y(0) = 1$.

(A) Graph the approximate solution and the slope field for $0 \le x \le 1$ using Euler's method with a step size of $\Delta x = .1$ using [0, 1].1 by [0, 3].1. Use trace and fill in the table below.

(B) Graph the approximate solution and the slope field for $0 \le x \le 1$ using Runge-Kutta method with a step size of $\Delta x = .1$ on the same set of coordinate axes. Use trace and fill in the table below.

(C) The exact solution of this equation is $y = 5e^x - 7x - 4$. Fill in the table below, using three-decimal place accuracy, to compare the numerical solutions and the exact solution.

(D) Graph the exact solution on the same set of coordinate axes.

x	Euler's Method	Runge-Kutta Method	Exact Solution
0			
0.1			
0.2			
0.3			
0.4			
0.5			
0.6			
0.7			
0.8			
0.9			
1.0			

Solution (A) and (C):

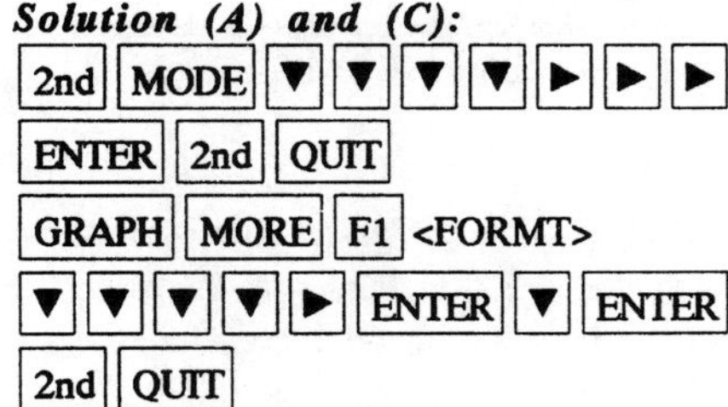

Set the calculator in DifEq mode.

Return to the Home Screen.

Set the calculator in Euler mode with the slope field turned on

Return to the Home Screen.

Turn all the statistical plots off by pressing: 2nd STAT F3 <PLOT> F5 <PlOff>

GRAPH

Q is used in place of y , and t is used in place of x in the differential equations mode. Change y to Q and x to t in the differential equation. We now have: Q'(t)=Q+7t-3. Up to nine differential equations can be entered in the list. We will use Q1. Enter the differential equation into the equation editor as Q1+7t-3.

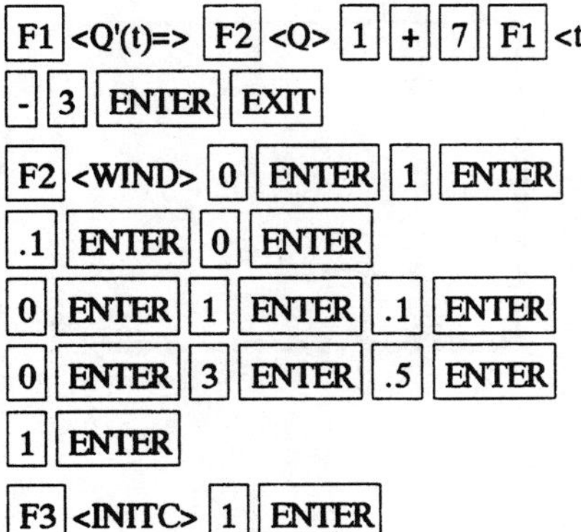

Set the step parameters and the window variables. $0 \le t \le 1$ with a step size of .1. This means tMax = 1 and tStep = .1.

The window variables can be set at [0, 1].1 by [0, 3].5.

Set the EStep to 1.

Set the initial conditions at tMin=0, tMax=1, and tStep=.1.

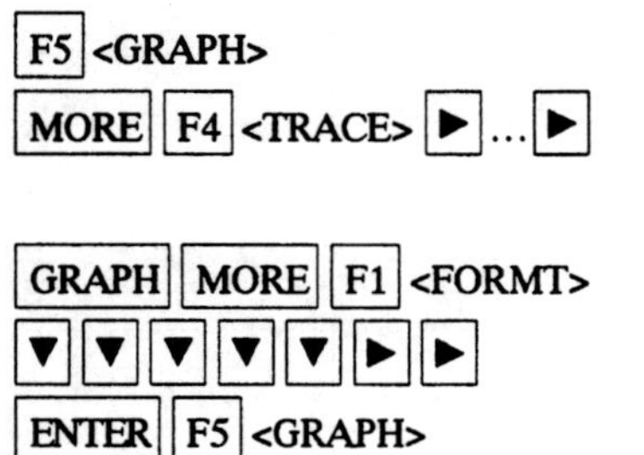

F5 <GRAPH>

Get the graph.

MORE F4 <TRACE> ▶ ... ▶

Use the arrow keys and read the numbers of the bottom of the screen. Record them in the table in Part (C).

GRAPH MORE F1 <FORMT>
▼ ▼ ▼ ▼ ▼ ▶ ▶
ENTER F5 <GRAPH>

The graph shown below does not have the slope field displayed. To turn the slope field display off, select f1 <FORMT> on the differential equation graph menu, arrow down to the bottom line, arrow to the right to select FldOff, and press enter.

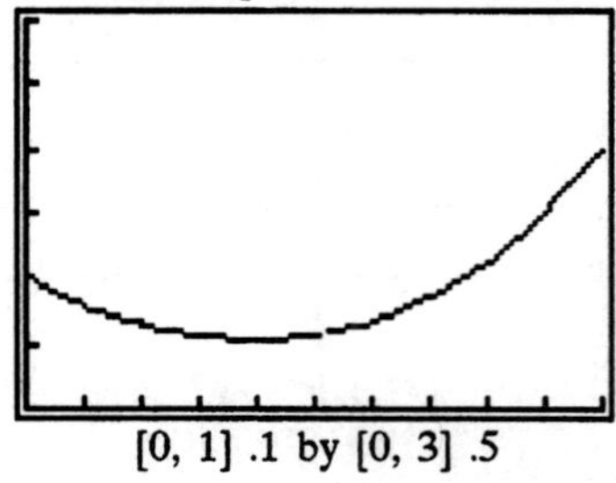

[0, 1] .1 by [0, 3] .5

Solution (B) and (C):

GRAPH MORE F1 <FORMT>
▼ ▼ ▼ ▼ ENTER
F5 <GRAPH>

This is the same as Part (A) but with the calculator to RK method with the slope field turned off. All we need to do is to change from the Euler mode to the RK mode. Repeat tracing the graph and record the results in the table.

You cannot display two graphical solutions at one time on the calculator.

To compare graphs, plot by hand on a piece of paper.

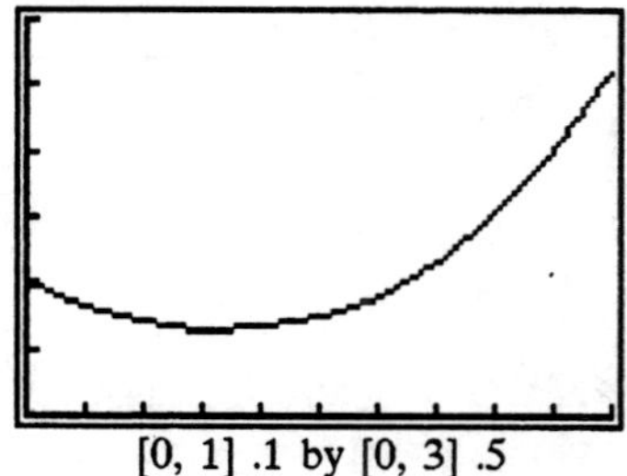

[0, 1] .1 by [0, 3] .5

Solution (C):

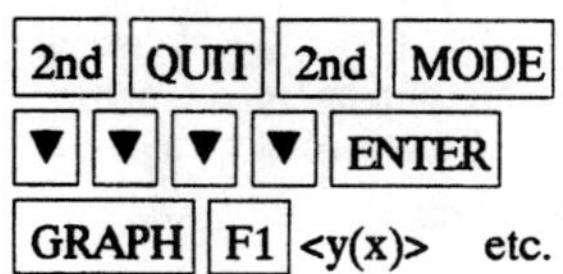

Change the back to function mode.

Store the exact solution in the calculator and evaluate it for the values of x in the table. See Section D-4 Example 3 Solution (B).

To compare the three graph, plot by hand on a piece of paper. They should look like:

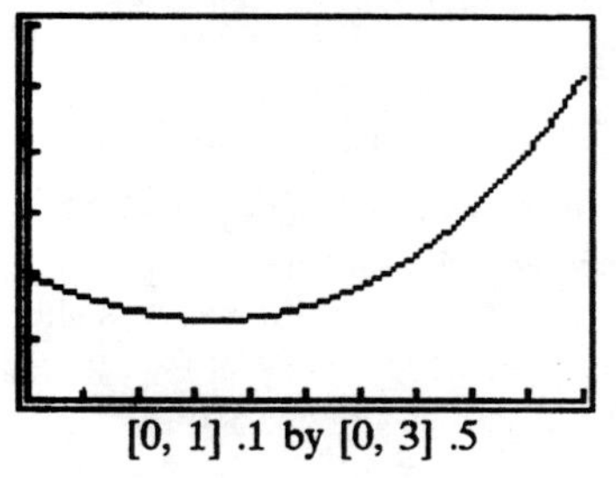

Runge-Kutta

[0, 1] .1 by [0, 3] .5

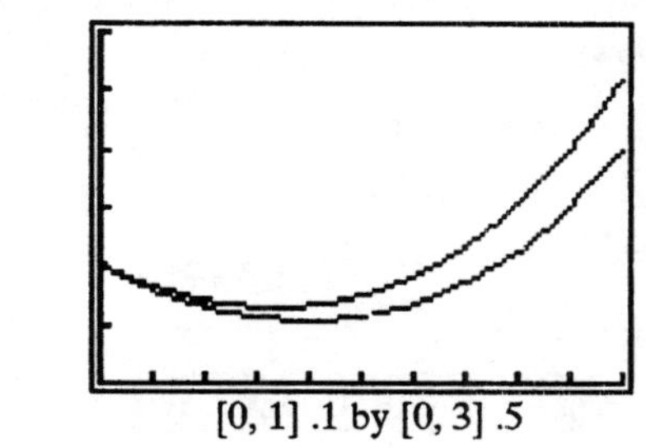

Exact

Euler

[0, 1] .1 by [0, 3] .5

Notice that the exact solution and the approximate solution found by the Runge-Kutta method are almost identical. The Runge-Kutta yields a more accurate approximation, but is not as fast to perform on the calculator.

x	Euler's Method	Runge-Kutta Method	Exact Solution
0	1	1	1
0.1	.800	.826	.826
0.2	.650	.707	.707
0.3	.555	.649	.649
0.4	.521	.659	.659
0.5	.553	.743	.744
0.6	.657	.910	.911
0.7	.844	1.168	1.169
0.8	1.118	1.527	1.528
0.9	1.490	1.997	1.998
1.0	1.969	2.590	2.591

D-16 Statistics

EXAMPLE 1: Given the following data, find: (A) the mean, (B) the sample standard deviation, (C) a histogram with first class beginning at 10 and width 4, and (D) repeat Parts A and B after deleting the data value 89 from the list. Use two decimal place accuracy. How did eliminating this outlier effect the mean and standard deviation?

23	56	73	12	89	52	45
38	18	21	15	37	42	51
19	40	57	38	54	60	39

Solution (A) and (B):

2nd QUIT

Return to the Home Screen.

2nd STAT F2 <EDIT>

▲ CLEAR ▼

Get the statistics menu. Clear the statistical registers by placing the cursor on the name of the list at the top of the screen by using the up arrow key and pressing CLEAR. Return to the list using the down arrow key and enter the data as *xStat*.

23 ENTER 56 ENTER … 39 ENTER

Use the down arrow key ▼ or press ENTER to move down one line in the list.

If you make a mistake when entering data, use the arrow keys to highlight the value and type it in again.

EXIT to exit the edit mode.

To remove a data value, use the arrow keys to get the data value and press DEL.

2nd STAT F1 <CALC> F1 <OneVa>

2nd ALPHA x ALPHA S

2nd ALPHA T 2nd ALPHA A

2nd ALPHA T ENTER

Get the calculate menu and choose OneVa, one-variable, calculations. Enter the name of the list as xStat. This can be found on the CATLG-VARS menu or you can type it in using the ALPHA key and the letters.

The mean is 41.86 and the sample standard deviation is 19.82.

Solution (C):

2nd QUIT

Return to the Home Screen.

2nd STAT F3 <PLOT> F1 <PLOT1>

ENTER ▼ F4 <HIST> ▼ F1 <xSTAT>

EXIT EXIT

Set up the statistical plot to plot a histogram.

GRAPH F1 <y(x)=>

Clear or deselect functions stored in the function list.

F5 <SELCT> EXIT

This will deselect the function the cursor is on. The = sign will no longer be highlighted. Deselect all functions stored or delete them using CLEAR.

MORE F2 <DRAW>

MORE MORE F1 <CLDRW> EXIT

Clear the drawings.

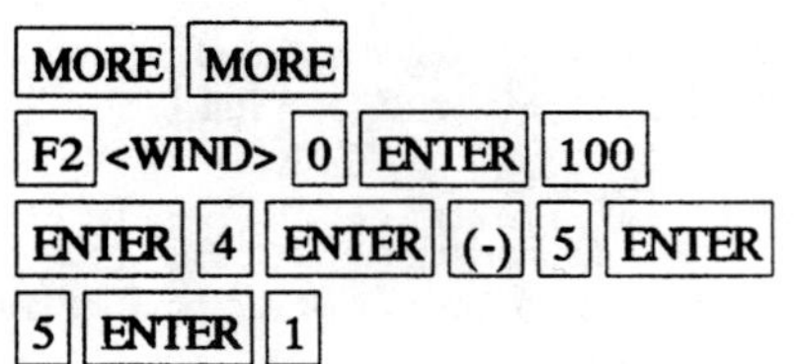

Set the window variables.
The smallest value is 12 so let xMin=0.
The greatest value is 89. Let us set xMax=100. Let's use xScl=4. This sets the histogram class width.
Let yMin=-5 to get some room below the histogram. We do not know the frequencies yet so let yMax=5 for a first try. Let yScl=1.

F5 <GRAPH>

Draw the histogram.

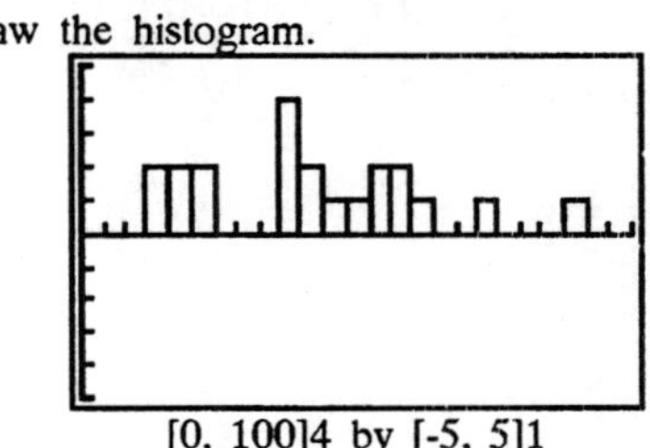

[0, 100]4 by [-5, 5]1

Solution (D):

2nd QUIT — Return to the Home Screen.

2nd STAT F2 <EDIT> ▼ ... ▼ — Get the data list and arrow down to the 89.

DEL EXIT — Delete the item.

2nd STAT F1 <CALC> F1 <OneVa> — Get the statistics menu and the one variable sample statistics.

2nd ALPHA x ALPHA S
2nd ALPHA T 2nd ALPHA A
2nd ALPHA T ENTER — The mean is now 39.5 and the sample standard deviation is 17.05. Notice how much influence an outlier has!

EXAMPLE 2: Given the following bivariate data: (A) find the correlation coefficient, (B) find the least squares line, (C) graph the least squares line, (D) plot the scatter diagram, and (E) sort the data on x and join the points using straight line segments.

x	5	8	10	29	9	24
y	3	4	8	20	3	19

Solution (A) & (B):

2nd QUIT — Return to the Home Screen.

2nd STAT F2 <EDIT> — Get the statistics menu.

▲ CLEAR ▼ 5 ENTER 8 ENTER ... — Clear the statistical register for xStat and enter the data.

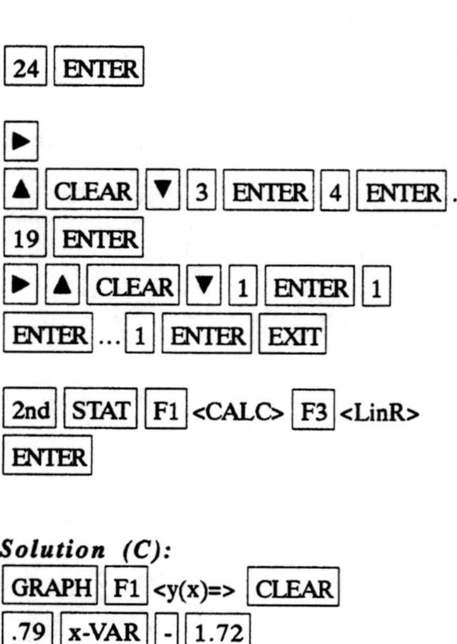

Clear the statistical register for yStat and enter the data. Move to the fStat list.

Move to the yStat list.

Clear the statistical register for fStat and enter 1's for the frequency of the pairs of data (xStat, yStat).

Linear regression correlation coefficient: corr = .98.
a=-1.72 and b=.79, so the least squares line is $\hat{y}$ = .79x - 1.72.

Solution (C):

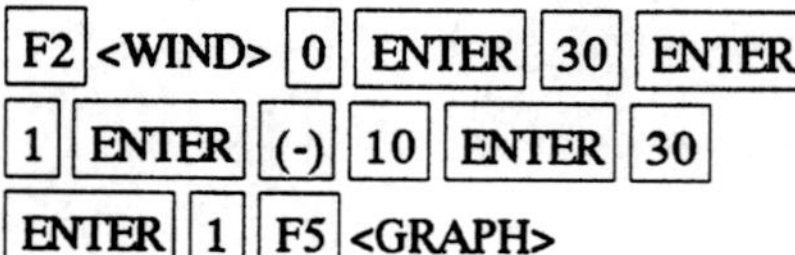

Clear or deselect all functions stored in the function list.
Set the cursor at $y1$ and enter the equation of the regression line.

If any of the plots are highlighted at the top of the screen, move the cursor up using the arrow keys and press enter to deselect them (to turn them off).

F2 <WIND> 0 ENTER 30 ENTER
1 ENTER (-) 10 ENTER 30
ENTER 1 F5 <GRAPH>

Set the window variables. Observe the data to determine what the dimensions of the viewing window should be.

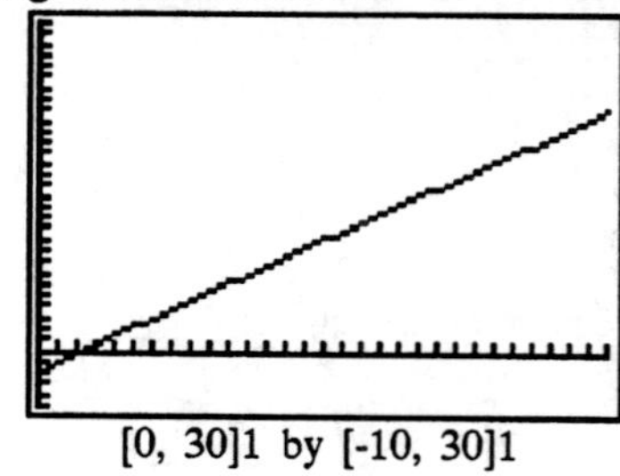

[0, 30]1 by [-10, 30]1

Solution (D):

Keystrokes	Explanation
2nd QUIT	Return to the Home Screen.
2nd STAT	Set up the statistical plot for a scatter diagram.
F3 <PLOT> F1 <PLOT1>	
ENTER ▼ F1 <SCAT> ▼ F1 <xStat>	
▼ F2 <yStat> ▼ F1 <□>	
EXIT EXIT	
GRAPH F1 <y(x)=> F5 <SELCT> EXIT	Clear or deselect functions stored in the function list except the equation of the regression line.
F2 <WIND> 0	Set the window variables. Observe the data to determine the dimensions of the viewing window should be. Set the y value
ENTER 30 ENTER 1	less than the minimum value so if you trace,
(-) 10 ENTER 30 ENTER 1	the coordinates at the bottom of the screen do not cover up the graph.
F5 <GRAPH>	

[0, 30]1 by [-10, 30]1

Solution (E):

Keystrokes	Explanation
2nd QUIT	Return to the Home Screen.
2nd QUIT 2nd LIST F5 <OPS>	Sort the data on x variable.
MORE MORE F5 <Sortx> ENTER	
EXIT EXIT	
GRAPH F1 <y(x)=> CLEAR	Clear or deselect all functions in the function list.
EXIT EXIT	
2nd STAT F4 <DRAW>	Draw the line that joins the points in the order they are listed in the xStat list.
F3 <xyLINE>	The points will be joined in the order of the x data in the data list.

[CLEAR]

Remove the menu bars.

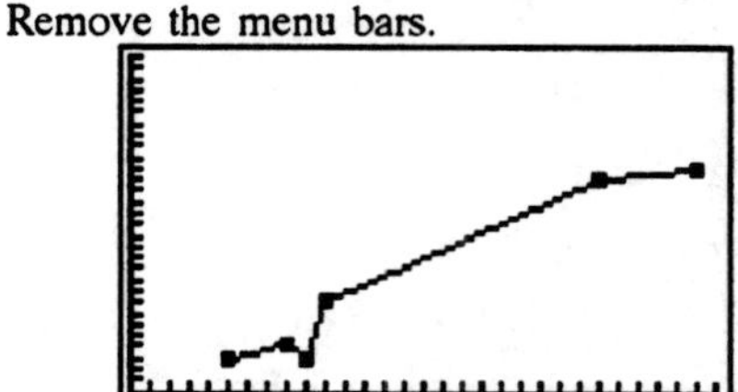

[0, 30]1 by [-10, 30]1

If the Plot is turned on, the points will also be plotted (see Part (D) above).

D-17 Setting the Display Mode for Numbers

Numbers can be entered into the calculator in scientific notation, a set number of significant digits, or a fixed number of decimal places.

EXAMPLE 1: Calculate $(9.35417\times 10^4)(-3.64389\times 10^{-6})$ using:
(A) scientific notation, (B) two significant digits, and (C) Four decimal place accuracy.

Solution (A):

[2nd] [MODE] [▶] [ENTER]
[2nd] [QUIT] [9.35417] [EE] [4] [×]
[(-)] [3.64389] [EE] [(-)] [6] [ENTER]

The calculator is set to scientific mode with floating decimals. The result displayed on the calculator is -3.40855665213E-1.

Solution (B):

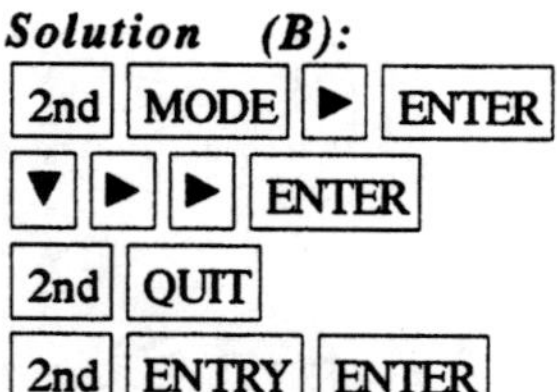

[2nd] [MODE] [▶] [ENTER]
[▼] [▶] [▶] [ENTER]

Set the calculator to scientific notation with 1 digit instead of floating point.

[2nd] [QUIT]

Return to the Home Screen.

[2nd] [ENTRY] [ENTER]

Use the replay key to replay the calculation just done in Solution (A). The result to two significant digits is -3.4E-1.

Solution (C):

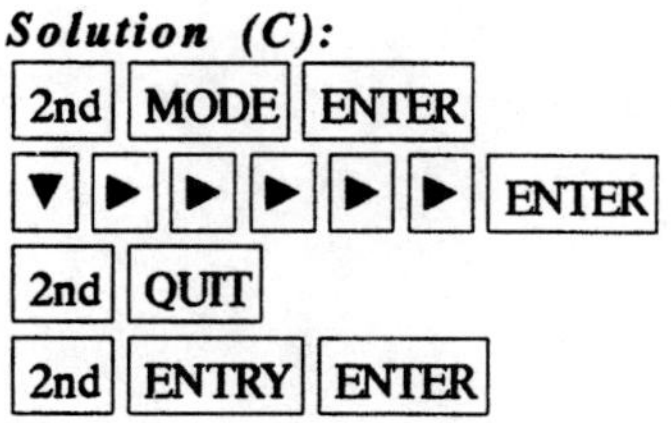

Set the calculator to normal mode with 4 digits.

Return to the Home Screen.

Use the replay key to replay the last calculation. The result to four decimal places is -.3409.

Change the calculator back to floating point. The examples in this manual are done using this setting unless otherwise noted.

NOTES

NOTES

Answers to Selected Problems

Chapter 1 A Beginning Library of Elementary Functions

1. The domain is $\{x \mid x \leq -1, x \geq 1\}$.

3. -5.25, -5.09, -5.0099, -5.000999. The value of $Q(x)$ gets close to -5 as x gets close to 2.

5.

h	.2	.1	.01	.001	→	0
$\frac{f(1.5+h) - f(1.5)}{h}$	.9	.8	.71	.701	→	.7

7.

x	$p(x)$	$p(x)$-p
2	1118	8
5	1010	20
10	830	15
14	686	9
21	434	1

9. (A)

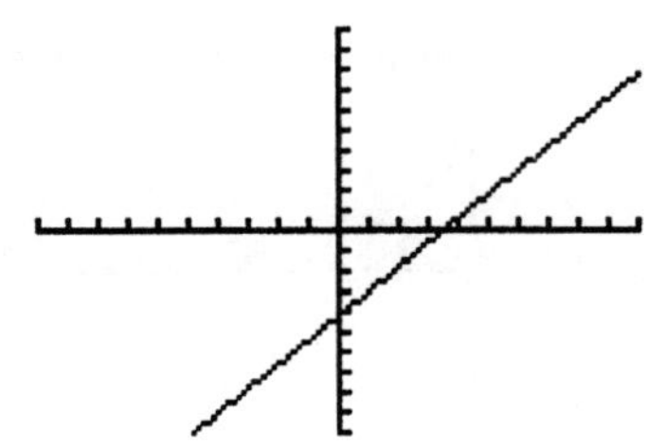

(B) & (C) x intercept is 3.5, y intercept is -4.2.

(D) $\{x \mid x > 3.5\}$

(E) The slope of the line is greater but the y intercept remains the same. Hence the x intercept of $g(x) = 2.2x - 4.2$ would be less than that of $f(x) = 1.2x - 4.2$. In fact, the x intercept of $g(x)$ is 1.9.

11. Yes.

13. Point of discontinuity at $x = 0$

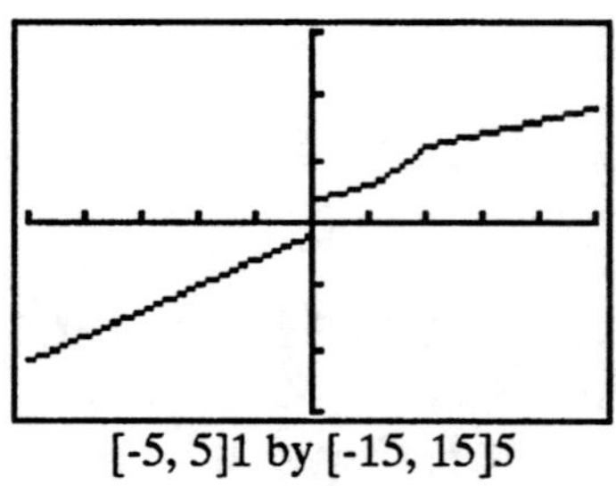

[-5, 5]1 by [-15, 15]5

15. The two lines are almost parallel. Compare the slopes 5.23/8.92 and 5.21/8.89.

17. The second graph is stretched vertically. All y values of the second graph are three times those of the first graph.
    The third graph is stretched vertically by a factor of 1.5 and reflected about the x axis. All y values of the first graph are multiplied by -1.5 to obtain the third graph.

Chapter 2 Additional Elementary Functions

1. -.84, 1.26, 11.29

3. $E(x) = .02x^3 - .10x^2 + 3x + 26.80$; $E(35) = \$1072.8$ billion

5. The vertical asymptote is $x = 1$.
   The horizontal asymptote is $y = 2$.

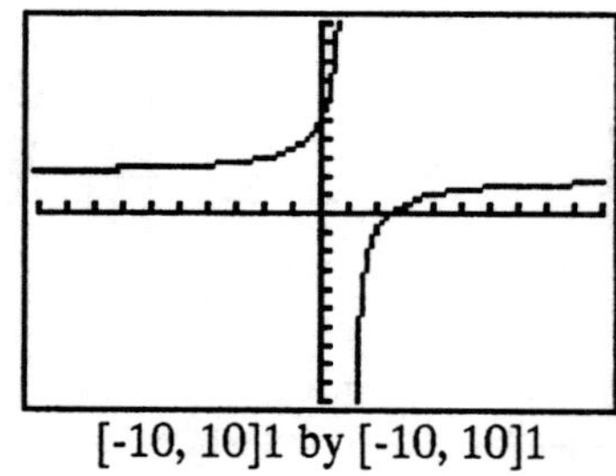

[-10, 10]1 by [-10, 10]1

7.

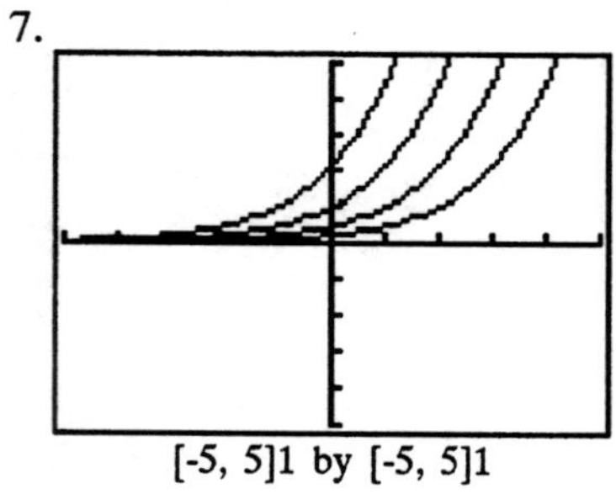

[-5, 5]1 by [-5, 5]1

Adding a constant to the x shifts the graph to the left.
Subtracting a constant to the x shifts the graph to the right.

9. 22.52825

11.

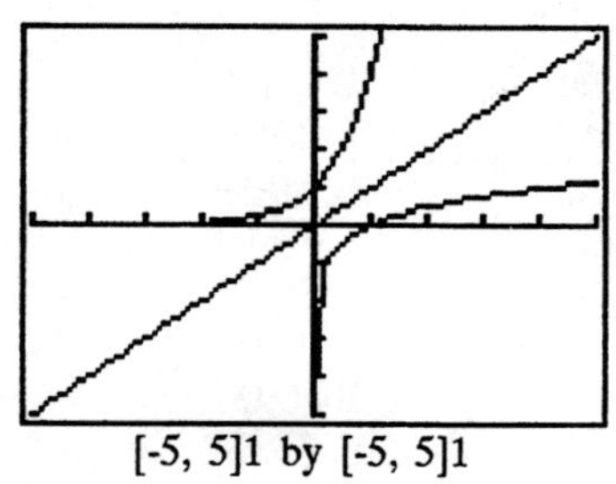

[-5, 5]1 by [-5, 5]1

Use the change-of-base formula to graph the inverse function.

Inverse of $y = 4^x$ is $y = \log_4 x = \dfrac{\ln(x)}{\ln(4)}$.

Chapter 3 Mathematics of Finance

1. 2, 175.70, 4351.41, 10878.52

3.

Quarter Number	Principal at Start of Quarter	Amount at End of Quarter	Interest Earned During the Quarter
1	6925.00	7037.53	112.53
2	7037.53	7151.89	114.36
3	7151.89	7268.11	116.22
4	7268.11	7386.22	118.11
5	7386.22	7506.24	120.03
6	7506.24	7628.22	121.98

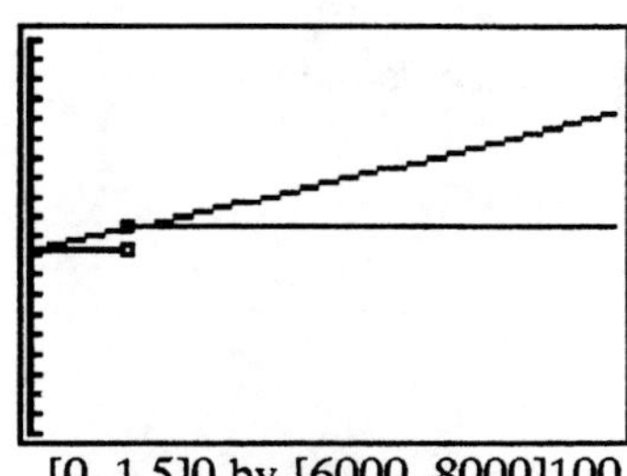

[0, 1.5]0 by [6000, 8000]100

5. S = \$1256.85, A = \$1318.52; Total interest earned = \$75.05.

7.

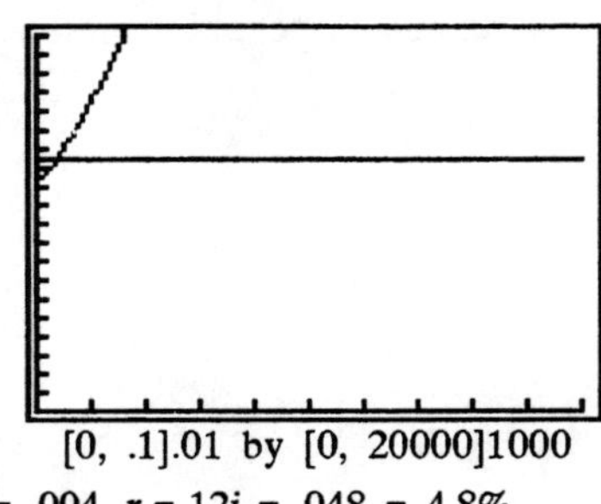

[0, .1].01 by [0, 20000]1000

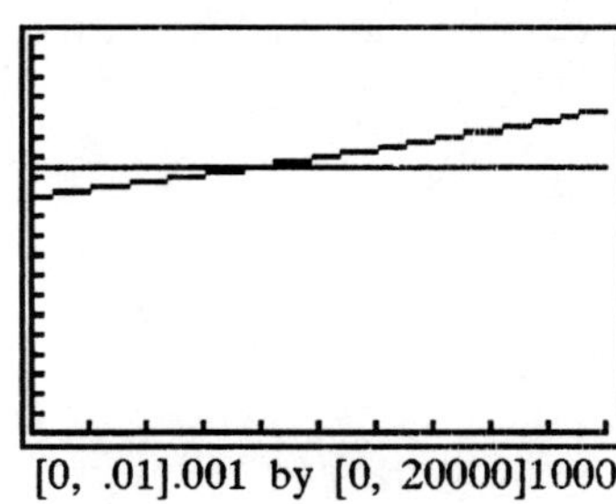

[0, .01].001 by [0, 20000]1000

$i = .004$, $r = 12i = .048 = 4.8\%$.

9. The payment is $307.38. The total interest paid is $188.56.

Period Number	Payment	Interest	Balance Reduction	Unpaid Balance
1	307.38	28.58	278.80	3221.20
2	307.38	26.31	281.07	2940.13
3	307.38	24.01	283.37	2656.76
4	307.38	21.70	285.68	2371.08
5	307.38	19.36	288.02	2083.06
6	307.38	17.01	290.37	1792.69
7	307.38	14.64	292.74	1499.95
8	307.38	12.25	295.13	1204.82
9	307.38	9.84	297.54	907.28
10	307.38	7.41	299.97	607.31
11	307.38	4.96	302.42	304.89
12	307.38	2.49	304.89	0.00

Chapter 4 Systems of Linear Equations; Matrices

1. The lines appear to intersect in a single point on the graph using [-10,10]1 by [-10,10]1. Graph using viewing window variables of [.4, .5].01 by [-2.5, -2.0].1 to see that the lines do not intersect in a single point. Hence there is no solution.

3. There will still be no solution. Changing the constant to 14 shifts the line downward 2 units.

5. $x = .29, y = -2.85, z = -3.65$

7. $x = 1.04, y = -.95, z = 1.07$

9. $\begin{bmatrix} \$54 & \$48 \\ \$36 & \$32 \end{bmatrix}$ and $\begin{bmatrix} \$60.53 & \$48 \\ \$40.35 & \$32 \end{bmatrix}$

   The labor costs increase $6.53 per ski for the trick ski and $4.35 per ski for the slalom ski.

Chapter 5 Linear Inequalities and Linear Programming

1. Corner points are (0, 7.14), (0, 10.06), (6.13, 0), (26.83, 0), and (2.66, 1.45)

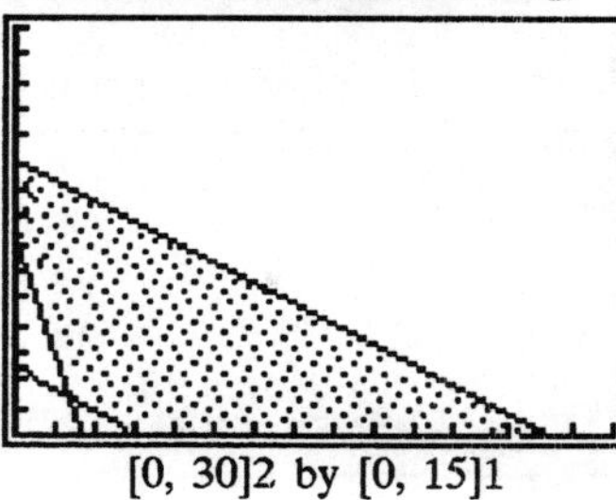

[0, 30]2 by [0, 15]1

3. Maximum at (26.83, 0), minimum at (0, 7.14).

5. A maximum profit of $2520 is achieved if 28 standard tents and 0 expedition tents are manufactured and sold. Substituting the solution into the slack variable equations shows that $s_1 = 4$ and s_2 is zero. This means that there are 4 hours in the cutting department that are not used. All available hours in the assembly department are used.

Chapter 6 Probability

1. (A) 2,496,960 (B) 6,991,488 (C) 621,075 (D) 1,366,365

3. Using a random number generator seed of 10 and entering [IPart] (25 x [RAND]) we get: 11, 13, 6, 17, and 21. Note that we use 25 as the number in the command since we want to generate numbers x such that $0 \le x \le 24$.

5. (A) $\begin{bmatrix} .294 & .130 & .015 \\ .106 & .083 & .092 \\ .076 & .065 & .139 \end{bmatrix}$ (B) .618

   (C) P(H) =.476; the events are not independent.

7. (A) (B) .25

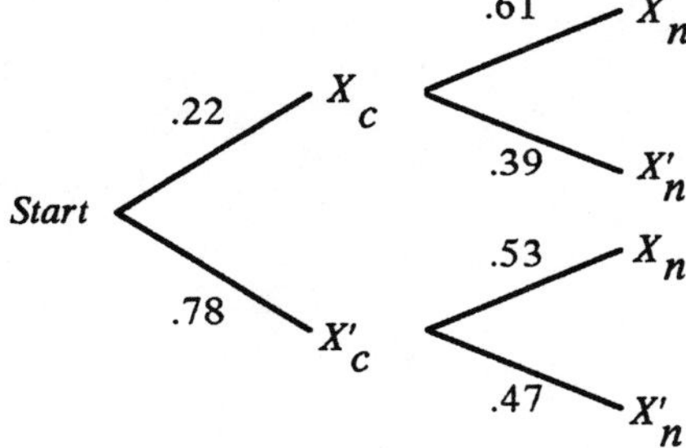

9. (A) .1

   (B) In the long run this player will win $.10 each game on the average. Numbers 5 and 6 are both the best numbers for the player to choose since they have the highest probability.

Chapter 7 Data Description and Probability Distributions

1. 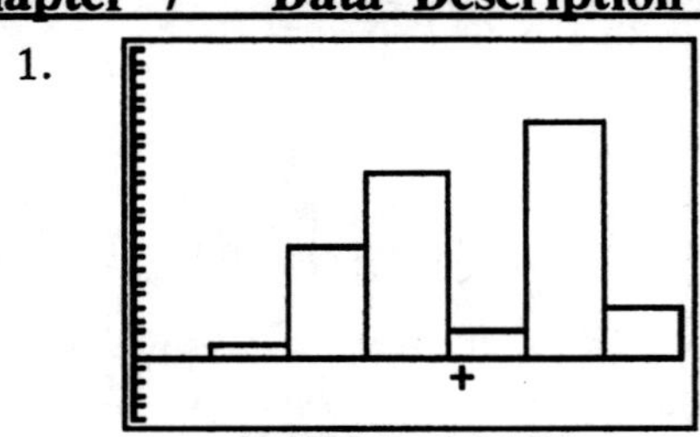

[0,35]5 by [-5,25]1

3. $\bar{x} = 21.12$, $s = 7.07$.

5. P(47)+P(46)+P(45) = .422 chance that 3, 4, or 5 will not recover.

Chapter 8 Games and Decisions

1. 1.17 This means that R is expected to win $1.17 per game in the long run.

3. $P^* = [.604 \quad .396]$ $Q^* = \begin{bmatrix} .597 \\ .403 \end{bmatrix}$ The game is not fair since $P^*MQ^* = .824 \neq 0$.

Chapter 9 Markov Chains

1. (A)

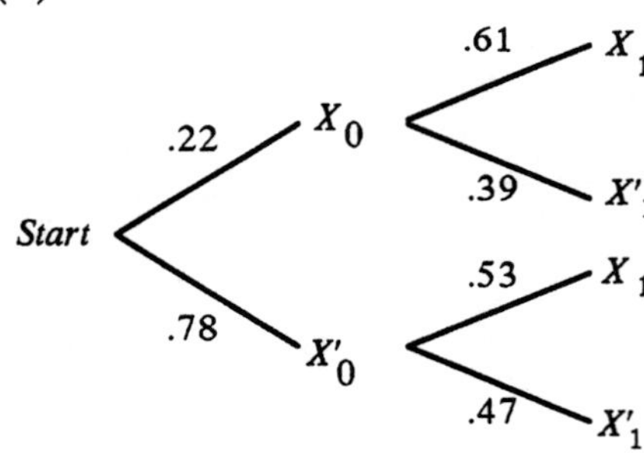

(B) Use Bayes formula.

$$P(X_0 \mid X_1) = \frac{P(X_0 \text{ and } X_1)}{P(X_1)}$$

$$= \frac{P(X_0 \text{ and } X_1)}{P(X_0 \text{ and } X_1) + P(X_0' \text{ and } X_1)}$$

$$= \frac{.22\text{x}.61}{.22\text{x}.61 + .78\text{x}.53} = .245$$

3. Transition matrix is $\begin{array}{c} \\ X_0 \\ X_0' \end{array} \begin{array}{c} \begin{array}{cc} X_1 & X_1' \end{array} \\ \begin{bmatrix} .61 & .39 \\ .53 & .47 \end{bmatrix} \end{array}$. The initial state is [.22 .78].

   (A) After five purchases, the company will have 58% of the market.

   (B) In the long run, the company will have 58% of the market.

5. (A) After five purchases, Product X will have 35% of the market and Product Y will have 29% of the market.

   (B) In the long run, Product X will have 35% of the market and Product Y will have 30% of the market.

NOTES

NOTES